Processing and Storage of Foods

Processing and Storage of Foods

Dharmesh Kuma

Processing and Storage of Foods

ISBN 978-93-5111-515-1

Published in 2015 in India by

RANDOM PUBLICATIONS

4376-A/4B, Gali Murari Lal, Ansari Road
New Delhi-110 002
Phone : +9111-43580356, 011-23289044, 011-43142548
e-mail: sales@randompublications.com,
info@randompublications.com, randomexports@gmail.com

Reprint 2023

Type Setting by : Friends Media, Delhi-110089
Printed at : Mehra Printers, Delhi-110 092

Preface

Food processing is a scientific and technological activity which covers a broader area than food preparation and cooking. It involves the application of scientific principles to slow down or stop the natural processes of food decay caused by micro-organisms, enzymes in the food, or environmental factors such as heat and sunlight, and so preserve the food. Food processing also uses science together with the creative imagination of the processor to change the eating quality of foods and provide people with interesting variety in their diets.

The reasons for processing and storage of foods may vary. In some countries the main aim is to preserve basic foods such as cereals, root crops or vegetables against periods of shortage; i.e. to increase the food security of populations throughout the year. Others wish to process foods to create employment and to generate additional income, either as a little extra money to supplement family incomes or to establish and expand a recognized food processing business.

The level of interest in food processing and storage has increased dramatically in recent years. It is due in part to the promotion of income-generating activities to help increase incomes and employment in rural areas, and also in part to the success of agricultural development programmes which have produced food surpluses then requiring preservation and processing. As a result there has been a corresponding upsurge in enquiries about food processing, and especially about the availability of low cost, small scale equipment and where it can be found.

It is hoped that this book will form a useful addition to the relatively limited information available on processing and storage of foods and serve its purpose as a tool to assist the improvement in peoples' livelihoods throughout the developing world.

Author

Contents

1

Introduction

Food processing is the transformation of raw ingredients into food, or of food into other forms. Food processing typically takes clean, harvested crops or butchered animal products and uses these to produce attractive, marketable and often long shelf-life food products. Similar processes are used to produce animal feed

Every parent in the world knows how to process foods they do it every day when preparing meals to feed their families. But 'food processing' as a scientific and technological activity covers a broader area than food preparation and cooking. It involves the application of scientific principles to slow down or stop the natural processes of food decay caused by micro-organisms, enzymes in the food, or environmental factors such as heat and sunlight, and so preserve the food. Food processing also uses science together with the creative imagination of the processor to change the eating quality of foods and provide people with interesting variety in their diets.

So the reasons for processing foods may vary. In some countries the main aim is to preserve basic foods such as cereals, root crops or vegetables against periods of shortage; i.e. to increase the food security of populations throughout the year.

Others wish to process foods to create employment and to generate additional income, either as a little extra money to supplement family incomes or to establish and expand a recognized food processing business. Small-scale food producers often start by working from home using domestic equipment; they often have little money to invest in equipment and little access to credit. However they must be able to produce uniform quality foods under hygienic conditions.

Some people will see food processing as their main source of income. They are entrepreneurs who will take out a loan to buy specialized equipment and secure working capital and, if successful, they will develop business and marketing skills to expand and diversify their enterprise.

For those who earn an income from food processing there are special problems which make this type of business different to most others: many raw materials are highly perishable and will spoil quickly after harvest or slaughter unless properly processed. They may also be highly seasonal which means that they can only be processed for part of the year, or alternatively they are part-processed for intermediate storage until they are needed.

Added to this, foods are biological materials whose composition varies as a result of the actions of weather, pests and diseases. This can mean unpredictable supplies and costs for raw materials. Some processed foods also have a seasonal demand (such as for festivals and ceremonies), which further complicates the business of food processing.

Even after processing, foods do not keep indefinitely. The ‘shelf-life’ of processed foods can vary from a few days to several months or years. The distribution and sales methods used by the processor must be suited to the expected shelf-life of the food and carefully organized so that customers receive the food before it spoils.

Packaging is an important means of controlling the shelf-life of foods, but there are universal problems in finding suitable packaging materials in developing countries. This is one of the most important constraints on small-scale food processors. The technically advanced plastic films, cartons and cans usually have to be imported, and even if foreign exchange restrictions or available supplies permit, they can be very expensive. Traditional alternatives such as leaves, clay pots etc. do not perform as well technically and are often perceived by customers as inferior. This puts the business at a marketing disadvantage compared to equivalent imported products.

In all food processing activities there is the overriding concern to avoid food poisoning. Food is the only commodity that people buy every day and take into their bodies. Processors and processing methods must meet strict standards of cleanliness and production control to avoid the risk of harming or even killing their customers by allowing the growth of food poisoning organisms in their products.

In no other type of business do processors operate under these multiple complex technical constraints. However, they do share with other types of small business the difficulties of operating in a frequently hostile economic environment. In the majority of developing countries the bulk of food processing enterprises are on a small scale and are located in the informal sector. They are rarely formed into associations and have little economic power or ability to seek such assistance as may be available. They will often need intermediaries, such as extension agents, to guide them to appropriate solutions for their own individual problems.

The larger, formal food processing sector may receive government support in the form of subsidies, foreign exchange allowances, price stabilization or guarantees and access to specialist advice. In contrast, the small-scale informal sector has no political influence, despite its combined voting power, and is therefore subject to the vagaries of the national (and often international) economic climate.

History of Food Processing

Food processing dates back to the prehistoric ages when crude processing incorporated slaughtering, fermenting, sun drying, preserving with salt, and various types of cooking (such as roasting, smoking, steaming, and oven baking). Salt-preservation was especially common for foods that constituted warrior and sailors' diets, until the introduction of canning methods. Evidence for the existence of these methods can be found in the writings of the ancient Greek, Chaldean, Egyptian and Roman civilizations as well as archaeological evidence from Europe, North and South America and Asia. These tried and tested processing techniques remained essentially the same until the advent of the industrial revolution. Examples of ready-meals also exist from preindustrial revolution times such as the Cornish pasty and Haggis. During ancient times and today these are considered processing foods. Food processing has also helped create quick, nutritious meals to give to busy families.

Modern food processing technology in the 19th and 20th century was largely developed to serve military needs. In 1809 Nicolas Appert invented a vacuum bottling technique that would supply food for French troops, and this contributed to the development of tinning and then canning by Peter

Durand in 1810. Although initially expensive and somewhat hazardous due to the lead used in cans, canned goods would later become a staple around the world. Pasteurization, discovered by Louis Pasteur in 1862, was a significant advance in ensuring the micro-biological safety of food.

In the 20th century, World War II, the space race and the rising consumer society in developed countries (including the United States) contributed to the growth of food processing with such advances as spray drying, juice concentrates, freeze drying and the introduction of artificial sweeteners, colouring agents, and preservatives such as sodium benzoate. In the late 20th century products such as dried instant soups, reconstituted fruits and juices, and self cooking meals such as MRE food ration were developed.

In western Europe and North America, the second half of the 20th century witnessed a rise in the pursuit of convenience. Food processing companies marketed their products especially towards middle-class working wives and mothers. Frozen foods (often credited to Clarence Birdseye) found their success in sales of juice concentrates and "TV dinners". Processors utilised the perceived value of time to appeal to the postwar population, and this same appeal contributes to the success of convenience foods today.

Methods of Food Preservation

Preservation usually involves preventing the growth of bacteria, fungi (such as yeasts), and other micro-organisms (although some methods work by introducing benign bacteria, or fungi to the food), as well as retarding the oxidation of fats which cause rancidity. Food preservation can also include processes which inhibit visual deterioration, such as the enzymatic browning reaction in apples after they are cut, which can occur during food preparation.

Many processes designed to preserve food will involve a number of food preservation methods. Preserving fruit by turning it into jam, for example, involves boiling (to reduce the fruit's moisture content and to kill bacteria, yeasts, etc.), sugaring (to prevent their re-growth) and sealing within an airtight jar (to prevent recontamination). There are many traditional methods of preserving food that limit the energy inputs and reduce carbon footprint.

Maintaining or creating nutritional value, texture and flavour is an important aspect of food preservation, although, historically, some methods drastically altered the character of the food being preserved. In many cases these changes have now come to be seen as desirable qualities – cheese, yoghurt and pickled onions being common examples.

Drying

Drying is one of the most ancient food preservation techniques, which reduces water activity sufficiently to prevent bacterial growth.

Refrigeration

Refrigeration preserves food by slowing down the growth and reproduction of micro-organisms and the action of enzymes which cause food to rot. The introduction of commercial and domestic refrigerators drastically improved the diets of many in the Western world by allowing foods such as fresh fruit, salads and dairy products to be stored safely for longer periods, particularly during warm weather.

Freezing

Freezing is also one of the most commonly used processes commercially and domestically for preserving a very wide range of food including prepared food stuffs which would not have required freezing in their unprepared state. For example, potato waffles are stored in the freezer, but potatoes themselves require only a cool dark place to ensure many months' storage. Cold stores provide large volume, long-term storage for strategic food stocks held in case of national emergency in many countries.

Vacuum Packing

Vacuum-packing stores food in a vacuum environment, usually in an air-tight bag or bottle. The vacuum environment strips bacteria of oxygen needed for survival, slowing spoiling. Vacuum-packing is commonly used for storing nuts to reduce loss of flavor from oxidation.

Salt

Salting or curing draws moisture from the meat through a process of osmosis. Meat is cured with salt or sugar, or a combination of the two. Nitrates and nitrites are also often used to cure meat and contribute the characteristic pink color, as well as inhibition of Clostridium botulinum.

Sugar

Sugar is used to preserve fruits, either in syrup with fruit such as apples, pears, peaches, apricots, plums or in crystallized form where the preserved material is cooked in sugar to the point of crystallisation and the resultant product is then stored dry. This method is used for the skins of citrus fruit (candied peel), angelica and ginger. A modification of this process produces glacé fruit such as glacé cherries where the fruit is preserved in sugar but is then extracted from the syrup and sold, the preservation being maintained by the sugar content of the fruit and the superficial coating of syrup. The use of sugar is often combined with alcohol for preservation of luxury products such as fruit in brandy or other spirits. These should not be confused with fruit flavored spirits such as cherry brandy or Sloe gin.

Smoking

Smoking is used to lengthen the shelf life of perishable food items. This effect is achieved by exposing the food to smoke from burning plant materials such as wood. Most commonly subjected to this method of food preservation are meats and fish that have undergone curing. Fruits and vegetables like paprika, cheeses, spices, and ingredients for making drinks such as malt and tea leaves are also smoked, but mainly for cooking or flavoring them. It is one of the oldest food preservation methods, which probably arose after the development of cooking with fire.

Artificial Food Additives

Preservative food additives can be antimicrobial; which inhibit the growth of bacteria or fungi, including mold, or antioxidant; such as oxygen absorbers, which inhibit the oxidation of food constituents. Common antimicrobial preservatives include calcium propionate, sodium nitrate, sodium nitrite, sulfites (sulfur dioxide, sodium bisulfite, potassium hydrogen sulfite, etc.) and disodium EDTA. Antioxidants include BHA and BHT. Other preservatives include formaldehyde (usually in solution), glutaraldehyde (kills insects), ethanol and methylchloroisothiazolinone.

Pickling

Pickling is a method of preserving food in an edible anti-microbial liquid. Pickling can be broadly categorized into two categories: chemical pickling and fermentation pickling.

In chemical pickling, the food is placed in an edible liquid that inhibits or kills bacteria and other micro-organisms. Typical pickling agents include brine (high in salt), vinegar, alcohol, and vegetable oil, especially olive oil but also many other oils. Many chemical pickling processes also involve heating or boiling so that the food being preserved becomes saturated with the pickling agent. Common chemically pickled foods include cucumbers, peppers, corned beef, herring, and eggs, as well as mixed vegetables such as piccalilli.

In fermentation pickling, the food itself produces the preservation agent, typically by a process that produces lactic acid. Fermented pickles include sauerkraut, nukazuke, kimchi, surströmming, and curtido. Some pickled cucumbers are also fermented.

Lye

Sodium hydroxide (lye) makes food too alkaline for bacterial growth. Lye will saponify fats in the food, which will change its flavor and texture. Lutefisk uses lye in its preparation, as do some olive recipes. Modern recipes for century eggs also call for lye. Masa harina and hominy use agricultural lime in their preparation and this is often misheard as 'lye'.

Canning and bottling

Canning involves cooking food, sealing it in sterile cans or jars, and boiling the containers to kill or weaken any remaining bacteria as a form of sterilization. It was invented by Nicolas Appert. Foods have varying degrees of natural protection against spoilage and may require that the final step occur in a pressure cooker. High-acid fruits like strawberries require no preservatives to can and only a short boiling cycle, whereas marginal fruits such as tomatoes require longer boiling and addition of other acidic elements. Low acid foods, such as vegetables and meats require pressure canning. Food preserved by canning or bottling is at immediate risk of spoilage once the can or bottle has been opened.

Lack of quality control in the canning process may allow ingress of water or micro-organisms. Most such failures are rapidly detected as decomposition within the can causes gas production and the can will swell or burst. However, there have been examples of poor manufacture (underprocessing) and poor hygiene allowing contamination of canned food by the obligate anaerobe Clostridium botulinum, which produces an acute toxin within the food, leading to severe illness or death. This organism produces no gas or obvious taste and remains undetected by taste or smell. Its toxin is denatured by cooking, though. Cooked mushrooms, handled poorly and then canned, can support the growth of Staphylococcus aureus, which produces a toxin that is not destroyed by canning or subsequent reheating.

Jellying

Food may be preserved by cooking in a material that solidifies to form a gel. Such materials include gelatine, agar, maize flour and arrowroot flour. Some foods naturally form a protein gel when cooked such as eels and elvers, and sipunculid worms which are a delicacy in Xiamen in Fujian province of the People's Republic of China. Jellied eels are a delicacy in the East End of London where they are eaten with mashed potatoes. Potted meats in aspic, (a gel made from gelatine and clarified meat broth) were a common way of serving meat off-cuts in the UK until the 1950s. Many jugged meats are also jellied.

Jugging

Meat can be preserved by jugging, the process of stewing the meat (commonly game or fish) in a covered earthenware jug or casserole. The animal to be jugged is usually cut into pieces, placed into a tightly-sealed jug with brine or gravy, and stewed. Red wine and/or the animal's own blood is sometimes added to the cooking liquid. Jugging was a popular method of preserving meat up until the middle of the 20th century.

Irradiation

Irradiation of food is the exposure of food to ionizing radiation; either high-energy electrons or X-rays from accelerators, or by gamma rays (emitted from radioactive sources as Cobalt-60 or Caesium-137). The treatment has a range of effects, including killing bacteria, molds and insect pests, reducing the ripening and spoiling of fruits, and at higher doses inducing sterility. The technology may be compared to pasteurization; it is sometimes called 'cold pasteurization', as the product is not heated. Irradiation is not effective against viruses or prions, it cannot eliminate toxins already formed by microorganisms.

The radiation process is unrelated to nuclear energy, but it may use the radiation emitted from radioactive nuclides produced in nuclear reactors. Ionizing radiation is hazardous to life (hence its usefulness in sterilisation); for this reason irradiation facilities have a heavily shielded irradiation room where the process takes place. Radiation safety procedures ensure that neither the workers in such facility nor the environment receive any radiation dose from the facility. Irradiated food does not become radioactive, and national and international expert bodies have declared food irradiation as wholesome. However, the wholesomeness of consuming such food is disputed by opponents and consumer organizations. National and international expert bodies have declared food irradiation as 'wholesome'; UN-organizations as WHO and FAO are endorsing to use food irradiation. International legislation on whether food may be irradiated or not varies worldwide from no regulation to full banning. Irradiation may allow lower quality or contaminated foodstuffs to be rendered marketable.

It is estimated that about 500,000 tons of food items are irradiated per year worldwide in over 40 countries. These are mainly spices and condiments with an increasing segment of fresh fruit irradiated for fruit fly quarantine. food preservation is very good process

Pulsed Electric Field Processing

Pulsed electric field (PEF) processing is a method for processing cells by means of brief pulses of a strong electric field. PEF holds potential as a type of low temperature alternative pasteurization process for sterilizing food products. In PEF processing, a substance is placed between two electrodes, then the pulsed electric field is applied. The electric field enlarges the pores of the cell membranes which kills the cells and releases their contents. PEF for food processing is a developing technology still being researched. There have been limited industrial applications of PEF processing for the pasteurization of fruit juices.

Modified Atmosphere

Modifying atmosphere is a way to preserve food by operating on the atmosphere around it. Salad crops which are notoriously difficult to preserve are now being packaged in sealed bags with an atmosphere modified to reduce the oxygen (O_2) concentration and increase the carbon dioxide (CO_2) concentration. There is concern that although salad vegetables retain their appearance and texture in such conditions, this method of preservation may not retain nutrients, especially vitamins. Grains may be preserved using carbon dioxide by one of two methods; either using a block of dry ice placed in the bottom and the can is filled with grain or the container can be purged from the bottom by gaseous carbon dioxide from a cylinder or bulk supply vessel.

Carbon dioxide prevents insects, and depending on concentration, mold, and oxidation from damaging the grain. Grain stored in this way can remain edible for five years.

Nitrogen gas (N_2) at concentrations of 98% or higher is also used effectively to kill insects in grain through hypoxia. However, carbon dioxide has an advantage in this respect as it kills organisms through hypercarbia and depending on concentration hypoxia and, requiring concentrations of above 35%, or so. This makes carbon dioxide preferable for fumigation in situations where a hermetic seal cannot be maintained.

Air-tight storage of grains (sometimes called hermetic storage) relies on the respiration of grain, insects and fungi which can modify the enclosed atmosphere sufficiently to control insect pests. This is a method of great antiquity, as well as having modern equivalents. The success of the method relies on have the correct mix of sealing, grain moisture and temperature.

High Pressure Food Preservation

High pressure food preservation refers to high pressure used for food preservation. "Pressed inside a vessel exerting 70,000 pounds per square inch (480 MPa) or more, food can be processed so that it retains its fresh appearance, flavour, texture and nutrients while disabling harmful microorganisms and slowing spoilage. By 2005 the process was being used for products ranging from orange juice to guacamole to deli meats and widely sold.

Burial in the Ground

Burial of food can preserve it due to a variety of factors: lack of light, lack of oxygen, cool temperatures, pH level, or desiccants in the soil. Burial may be combined with other methods such as salting or fermentation. Most foods can be preserved in soil that is very dry and salty (thus a desiccant), or soil that is frozen.

Many root vegetables are very resistant to spoilage and require no other preservation than storage in cool dark conditions, for example by burial in the ground, such as in a storage clamp. Century eggs are created by placing eggs in alkaline mud (or other alkaline substance) resulting in their "inorganic" fermentation through raised pH instead of spoiling. The fermentation preserves them and breaks down some of the complex, less flavorful proteins and fats into simpler more flavorful ones. Cabbage was traditionally buried in the fall in northern farms in the USA for preservation. Some methods keep it crispy while other methods produce sauerkraut. A similar process is used in the traditional production of kimchi. Sometimes meat is buried under conditions which cause preservation. If buried on hot coals or ashes, the heat can kill pathogens, the dry ash can desiccate, and the earth can block oxygen and further contamination. If buried where the earth is very cold, the earth acts like a refrigerator.

Controlled Use of Micro-organism

Some foods, such as many cheeses, wines, and beers will keep for a long time because their production uses specific micro-organisms that combat spoilage from other less benign organisms. These micro-organisms keep pathogens in check by creating an environment toxic for themselves and other micro-organisms by producing acid or alcohol. Starter micro-organisms, salt, hops, controlled (usually cool) temperatures, controlled (usually low) levels of oxygen and/or other methods are used to create the specific controlled conditions that will support the desirable organisms that produce food fit for human consumption.

Biopreservation

Biopreservation is the use of natural or controlled microbiota or antimicrobials as a way of preserving food and extending its shelf life. Beneficial bacteria or the fermentation products produced by these bacteria are used in biopreservation to control spoilage and render pathogens inactive in food. It is a benign ecological approach which is gaining increasing attention.

Of special interest are lactic acid bacteria (LAB). Lactic acid bacteria have antagonistic properties which make them particularly useful as biopreservatives. When LABs compete for nutrients, their metabolites often include active antimicrobials such as lactic and acetic acid, hydrogen peroxide, and peptide bacteriocins. Some LABs produce the antimicrobial nisin which is a particularly effective preservative.

These days LAB bacteriocins are used as an integral part of hurdle technology. Using them in combination with other preservative techniques can effectively control spoilage bacteria and other pathogens, and can inhibiting the activities of a wide spectrum of organisms, including inherently resistant Gram-negative bacteria.

Hurdle Technology

Hurdle technology is a method of ensuring that pathogens in food products can be eliminated or controlled by combining more than one approach. These approaches can be thought of as "hurdles" the pathogen has to overcome if it is to remain active in the food. The right combination of hurdles can ensure all pathogens are eliminated or rendered harmless in the final product.

Hurdle technology has been defined by Leistner as an intelligent combination of hurdles which secures the microbial safety and stability as well as the organoleptic and nutritional quality and the economic viability of food products. The organoleptic quality of the food refers to its sensory properties, that is its look, taste, smell and texture.

Examples of hurdles in a food system are high temperature during processing, low temperature during storage, increasing the acidity, lowering the water activity or redox potential, or the presence of preservatives or biopreservatives. According to the type of pathogens and how risky they are, the intensity of the hurdles can be adjusted individually to meet consumer preferences in an economical way, without sacrificing the safety of the product.

Benefits and Drawbacks

Benefits

Benefits of food processing include toxin removal, preservation, easing marketing and distribution tasks, and increasing food consistency. In addition, it increases yearly availability of many foods, enables transportation of delicate perishable foods across long distances and makes many kinds of foods safe to eat by de-activating spoilage and pathogenic micro-organisms. Modern supermarkets would not exist without modern food processing techniques, long voyages would not be possible and military campaigns would be significantly more difficult and costly to execute.

Processed foods are usually less susceptible to early spoilage than fresh foods and are better suited for long distance transportation from the source to the consumer. When they were first introduced, some processed foods helped to alleviate food shortages and improved the overall nutrition of populations as it made many new foods available to the masses.

Processing can also reduce the incidence of food borne disease. Fresh materials, such as fresh produce and raw meats, are more likely to harbour pathogenic micro-organisms capable of causing serious illnesses.

The extremely varied modern diet is only truly possible on a wide scale because of food processing. Transportation of more exotic foods, as well as the elimination of much hard labour gives the modern eater easy access to a wide variety of food unimaginable to their ancestors.

Mass production of food is much cheaper overall than individual production of meals from raw ingredients. Therefore, a large profit potential exists for the manufacturers and suppliers of processed food products. Individuals may see a benefit in convenience, but rarely see any direct financial cost benefit in using processed food as compared to home preparation.

Processed food freed people from the large amount of time involved in preparing and cooking "natural" unprocessed foods. The increase in free time allows people much more choice in life style than previously allowed. In many families the adults are working away from home and therefore there is little time for the preparation of food based on fresh ingredients. The food industry offers products that fulfill many different needs: From peeled potatoes that only have to be boiled at home to fully prepared ready meals that can be heated up in the microwave oven within a few minutes.

Modern food processing also improves the quality of life for people with allergies, diabetics, and other people who cannot consume some common food elements. Food processing can also add extra nutrients such as vitamins.

Drawbacks

Any processing of food can effect its nutritional density, the amount of nutrients lost depending on the food and method of processing. Vitamin C, for example, is destroyed by heat and therefore canned fruits have a lower content of vitamin C than fresh ones. The USDA conducted a study in 2004, creating a nutrient retention table for several foods. A cursory glance of the table indicates that, in the majority of foods, processing reduces nutrients by a minimal amount. On average any given nutrient may be reduced by as little as 5%-20%.

Another safety concern in food processing is the use of food additives. The health risks of any additives will vary greatly from person to person;

for example sugar as an additive would be detrimental to those with diabetes. In the European Union, only food additives (e.g., sweeteners, preservatives, stabilizers) that have been approved as safe for human consumption by the European Food Safety Authority (EFSA) are allowed, at specified levels, for use in food products. Approved additives receive an E number (E for Europe), which at the same time simplifies communication about food additives in the list of ingredients across the different languages of the EU.

Food processing is typically a mechanical process that utilizes large mixing, grinding, chopping and emulsifying equipment in the production process. These processes inherently introduce a number of contamination risks. As a mixing bowl or grinder is used over time the food contact parts will tend to fail and fracture. This type of failure will introduce in to the product stream small to large metal contaminates. Further processing of these metal fragments will result in downstream equipment failure and the risk of ingestion by the consumer.

Food manufacturers utilize industrial metal detectors to detect and reject automatically any metal fragment. Large food processors will utilize many metal detectors within the processing stream to both ensure reduced damage to processing machinery as well risk to the consumer. The first industrial level metal detector pioneered by Goring Kerr was introduced back in 1947 for Mars Incorporated.

Performance Parameters for Food Processing

When designing processes for the food industry the following performance parameters may be taken into account:

— Hygiene, e.g. measured by number of micro-organisms per ml of finished product

— Energy efficiency measured e.g. by "ton of steam per ton of sugar produced"

— Minimization of waste, measured e.g. by "percentage of peeling loss during the peeling of potatoes'

— Labour used, measured e.g. by "number of working hours per ton of finished product"

— Minimization of cleaning stops measured e.g. by "number of hours between cleaning stops"

Problems often occur during preparation of batter mixes because flour and other powdered ingredients tend to form lumps or agglomerates as they are being mixed during production. A conventional mixer/agitator cannot break down these agglomerates, resulting in a lumpy batter. If lumpy batter is used to enrobe products, it causes an unsatisfactory appearance with misshapen or oversize products that do not fit properly into packaging. This can force production to a standstill. Furthermore batter mix is generally recirculated from an enrobing system back to a holding vessel; lumps then have a tendency to build up, reducing the flow of material and raising potential sanitation issues.

Using a high shear in-line mixer in place of a conventional agitator or mixer can quickly solve problems of agglomeration with dry ingredients. A single pass through a self-pumping, in-line mixer adds high shear to batter, which de-agglomerates the mix, resulting in a homogeneous, smooth batter. With a consistent, smooth batter, finished product appearance is improved; the effectiveness and hygiene of the recirculation system is increased; and a better yield of raw materials is achieved. By increasing overall product quality, the amount of raw materials needed is decreased, thereby lowering manufacturing costs.

High shear in-line mixers process food to be made faster and cheaper while increasing consistency of the finished food. Powder and liquid mixing systems are capable of rapidly incorporating large quantities of powders at high concentrations – agglomerate free and fully hydrated. Advances in technology have made processing equipment easy to clean, leading to a much safer processed food.

References

Alais, C. and G. Linden. 1991. *Food Biochemistry*. New York, Ellis Horwood Ltd., 222 pp.

Betschart, A. A. 1982. "World food and nutrition problems." *Cereal Food World* 27: 562.

Borgstrom, G. 1968. *Principals of Food Science*, Vol. 2. Food Microbiology and Biochemistry. New York, Macmillan.

Tucker, G.A. and L.F.J. Woods. 1995. *Enzymes in Food Processing.*Chapman and Hall, New York, 319 pp.

2

Food Processing for Improved Livelihoods

Food has been processed since communities first came together thousands of years ago. Most foods need some form of preparation and processing to make them more attractive to eat. Grains, vegetables, meats and fish are each unpalatable in their raw state, and some foods, such as cassava, are dangerous if eaten without processing. Even nuts, milk and fruits that are eaten raw can benefit from processing into a wide variety of other products.

Different types of processing have been developed over generations into the range of methods that we have today. In every region, country and even in individual villages, there are distinctive traditional processed foods that are well suited to the local climatic and socio-economic conditions (for example, the 2 000 different cheeses throughout the world, each with its own distinctive flavour and texture). In villages throughout the world, families inherit or develop specialist skills and become for example the village baker, brew!er or fish smoker. Traditional products have a high local demand and are often sought by people in other areas, so establishing trade and the development of local food businesses. In Sri Lanka, for example, some communities are known throughout the island for the quality of their buffalo curd (yoghurt), which is bought by traders and distributed over wide areas. Food preparation and processing therefore benefit communities by:

- increasing the variety of foods in the diet;
- creating special foods for cultural or religious occasions, thus reinforcing cultural identities;
- creating opportunities for sales and income generation.

However, processing does more than change the eating quality of raw foods. All foods are biological materials that begin to decay as soon as they are harvested or slaughtered. Processing slows down or stops this deterioration and thus allows foods to be preserved for extended periods. This benefits village communities in a number of ways.

— It enables food to be stored as a reserve against times of shortage to increase food security.

— It enables crops to be sold out of season when prices are higher.

Processing offers opportunities for villagers to diversify their sources of income. When farmers in an area grow similar crops, processing helps to avoid the effects of lowered prices and incomes when seasonal gluts occur at harvest time. Processing also enables farmers who grow low-value staple crops to add value and increase household incomes. For example, in many African countries processing sorghum into beer or processing cassava into gari or snack foods can form very successful small-scale businesses. In many Asian countries, value is similarly added to fruits and vegetables by processing them into a wide range of pickles, chutneys and other relishes. These small-scale operations are a major source of employment in rural villages, estimated at up to 60 percent of employment in some countries.

Many governments and international development agencies promote food processing as a means of alleviating poverty in rural areas. There are many advantages in choosing food processing over other income-generating activities.

— Food processing is accessible - everyone is familiar with the food that they grow and eat and, compared with some other types of business, there are fewer aspects to learn when getting started. Small-scale food processing is also particularly suitable for women, who may be the specific intended beneficiaries of development programmes.

— If chosen correctly, processed foods can have a good demand and offer the opportunity to generate good profits by adding value to raw materials.

— Crops or animals that are the raw materials for processing are usually readily available (and sometimes in surplus).

— Of the many different types of processing technologies, most are suitable for small-scale operation with an affordable level of investment by rural people.

— Domestic utensils can be used in many processes when starting up. When production expands, many types of processing equipment can be manufactured locally by metal workshops or carpenters, thus creating further employment.

— Most types of processing have few negative environmental effects.

There are two broad categories of food preparation and processing:

1. *Primary processing*, in which foods are stabilized after harvest and sometimes converted into a more convenient form for storage. Examples include drying crops, milling cereals and extracting cooking oils from oilseeds or nuts. These types of processing are described in more detail in the accompanying booklet in this series *High hopes for post-harvest*.
2. *Secondary processing*, in which fresh foods or the products of primary processing are made into a wide range of processed foods..

People process foods every day when preparing meals to feed their families. However, the term "food processing" is broader than preparing and cooking foods. It involves applying scientific and technological principles to preserve foods by slowing down or stopping the natural processes of decay. It also allows changes to the eating quality of foods to be made in a predictable and controlled way. Food processing uses the creative potential of the processor to change basic raw materials into a range of tasty attractive products that provide interesting variety in the diets of consumers.

All food manufacturers should make safe foods so that consumers are not at risk. This is not only microbiological risks, but also glass splinters, pesticides or other harmful materials that can get into the food and lower its quality. Consumers consider *eating quality* as the main factor when buying foods, and a food should fit in with traditional eating habits and cultural expectations of texture, flavour, taste colour and appearance. For some foods, *nutritional quality* (e.g. protein content, vitamins and minerals, etc.) is an important consideration. Product quality is affected by the raw materials, the processing conditions and the storage and handling that a food is subjected to after processing. Food processors should understand the composition of their foods because it enables them to predict the changes that take place during processing, the expected shelf life of the product and the types of microorganisms that can grow in it. This information is used to prevent food spoilage or food poisoning. Details of the composition of raw materials or products can be obtained from university food science departments, bureaux of standards or food research institutes.

Without processing, as much as 50 to 60 percent of fresh food can be lost between harvest and consumption. This may be due to inadequate storage facilities, which allow micro-organisms or pests to spoil the stored food. Improved storage can greatly reduce these losses. Processing methods that are suitable for village scale processing can be grouped into six categories (Table 1).

Table 1. Types of village food processing

Category of process	*Examples of types of processes*
Heating to destroy enzymes and micro-organisms.	Boiling, blanching, roasting, grilling, pasteurization, baking, smoking
Removing water from the food	Drying, concentrating by boiling, filtering, pressing
Removing heat from the food	Cooling, chilling, freezing
Increasing acidity of foods	Fermentation, adding citric acid or vinegar
Using chemicals to prevent enzyme and microbial activity	Salting, syruping, smoking, adding chemical preservatives such as sodium metabisulphite or sodium benzoate
Excluding air, light, moisture, micro-organisms and pests	Packaging

A number of other preparation methods (such as mixing, coating with batter, grinding, cutting, etc.) alter the eating quality of foods, but do not preserve them. It is important to note that the production of most food processed foods uses more than one of the categories in Table 1. For example, jam making involves heating, removing water, increasing the levels of acidity and sugar, and packaging. Smoking fish or meat involves heating, removing water and coating the surface with preservative smoke chemicals.

Effect of Processing on the Quality of Foods

In addition to preserving foods, secondary processing alters their eating quality. A good example is cereal grains, where primary processing by drying and milling produces flour, which remains inedible. Secondary processing is used to produce a wide range of bakery products, snack foods, beers and porridges, each having an attractive flavour, texture and/or colour. Eating quality is the main influence on whether customers buy a product. Foods that have an attractive appearance or colour are more likely to sell well and at a higher price. It is therefore in the interests of processing businesses to find out what it is that consumers like about a product using market assessments and ensure that the products meet their requirements. This is described below.

Scales of Operation

When operating as a business, food processing can take place at any scale from a single person upwards (Table 2). The focus of this booklet is on the smaller scales of operation from "home-scale" to "small-scale".

Table 2. Scales of commercial food processing

Scale of operation	*Characteristics*
Home- (or household-) scale	No employees, little or no capital investment
Micro- (or cottage-) scale	Less than 5 employees, capital investment less than US$1 000
Small-scale	5-15 employees, capital investment US$1 000-US$50 000
Medium-scale	16-50 employees, capital investment US$50 000-US$1 000 000
Large-scale	More than 50 employees, capital investment over US$1 000 000

Home-scale Processing

Foods that are intended for household consumption are usually processed by individual families or small groups of people working together. Many of the world's multinational food conglomerates started from a single person or family working from home. In developing countries, home-scale processors aim to generate extra income to meet family needs such as clothing or school fees. Where this is successful, many later expand production and develop first into a micro- or small-scale business, and later into larger scale operations.

Characteristically, home-scale processors cannot afford specialized food processing equipment and rely on domestic utensils, such as cooking pans and stoves for their production.

They may work part-time as the need for money arises and use part of the house, or an outbuilding for processing. However, in many situations the lack of dedicated production facilities means that there is a risk of contamination and product quality may be variable. This may reduce the value of the processed foods and the potential family income. A role of extension agents and training programmes is to upgrade facilities and hygiene, to introduce simple quality assurance techniques and improved packaging, to enable products to compete more effectively with those from larger processors.

Where families generate sufficient income from sales, some choose to invest in specialist equipment (such as a bakery oven, or a press for dewatering cassava or making cooking oil). In most cases, such equipment can be made by a competent local carpenter, bricklayer or blacksmith. This allows home-scale businesses to expand and become micro- or smallscale enterprises.

Micro-scale Processing

Whereas home processors may sell their products to neighbours or in village marketplaces, the move up to micro-scale processing requires additional skills and confidence to compete with other processors and to negotiate with professional buyers, such as retailers or middlemen. Similarly, although the quality of their products may be suitable for rural consumers, it may not be sufficient to compete with products from larger companies in other markets. To successfully expand to micro-scale production, village processors need technical skills to make consistently high quality products, and financial and marketing skills to make the business grow and become successful. They may require assistance to gain these skills and confidence, and short training programmes or technical extension workers can help them to establish improved production methods, quality assurance and selling techniques.

Small-scale Processing

The expansion to a small-scale processing operation requires additional investment to produce larger amounts of product in a dedicated production room. It is likely to require specialist equipment that is either made by a metal workshop in a nearby town or imported, because most rural blacksmiths do not have the necessary skills, equipment or materials to make such equipment. At this level of production, village processors are likely to be in competition with other small-scale businesses, larger companies and imported products. They need to develop attractive packaging, quality assurance techniques, and the financial and managerial skills needed to run a successful small business.

If the level of investment at this scale is too high for individual families, an alternative approach is for a group of people, such as a farmers' group, or a women's group, or a producer co-operative to operate the food processing business together. They invest jointly in the equipment and facilities, and market their products under a single brand name. There are many advantages to this approach including a greater willingness by lenders to make a loan if a group is sharing responsibility for the repayments, new employment opportunities for those without land, discouraging migration to larger towns or cities, and providing greater financial security and an improved standard of living to larger numbers of people.

Many governments promote the development of small-scale food processing enterprises because they:

- have the potential to create significant levels of employment;
- increase food security for growing urban populations as well as rural families;
- produce products that can substitute for imported foods or have export potential, and thus help reduce balance of payments problems and improve the overall prosperity of the country.

Food Processing for Sustainable Livelihoods

The normal diets of people in village communities throughout most of the world consist of a cereal or root crop staple food, which provides starchy carbohydrates, together with a source of animal or plant protein, such as meat, milk, fish, beans etc. Vitamins and minerals are provided by leafy vegetables, fruits or nuts, which are often made into a strong-tasting relish to accompany the relatively bland starchy staples. To remain healthy, people need to eat an adequate amount of food, but also ensure that the food contains sufficient amounts of specific nutrients.

Because most crops are seasonal, there are times of the year when either gluts can result in high levels of wastage or shortages can arise if adequate measures are not taken to preserve and store the foods. This is particularly important in areas that have a dry season or winter period when crops cannot be grown and animals are slaughtered because of a lack of fodder. In these situations stored dry grains or root crops provide energy; dried, salted or smoked meats, or cheeses provide a source of protein, vitamins and minerals; and processed fruits and vegetables such as pickles, chutneys or dried fruits or leaves provide vitamins and minerals. A few crops, including cassava and some types of beans also contain poisons or anti-nutritional components, which must be removed by processing to make the food safe to eat.

Where a household processes some of its crops or animals for sale, the additional diversified sources of income provide a greater degree of economic security. Other social and economic benefits that can arise from processing include: more efficient use of time and, labour when equipment is used to process crops; and enhanced social standing within the community for groups who run a successful processing operation. Programmes that promote food processing in rural communities can also be used to introduce a wide range of skills that are needed to improve the livelihoods of people in rural communities. Examples include greater confidence and negotiating skills, improved skills in managing income and expenditure, improved understanding of quality requirements of consumers and changes to crop production methods to meet consumer needs. Establishing a processing unit can offer alternative employment opportunities to young people who suffer from lack of access to land. It can also be used to introduce nutrition concepts to mothers' groups, or widen the career horizons of schoolchildren through operating a small food business. As the community is strengthened management skills, machinery repair workshops, market information networks, etc., can each be introduced. The village becomes more financially secure and improvements to schools, medical facilities and other services can follow.

Processing for Improved Nutrition

If the aim is to improve health and nutrition, an understanding of the current dietary problems is needed, together with any likely future changes to food consumption. A nutritional survey is conducted to identify deficiencies, and this is then used to decide which foods should be preserved to meet those deficiencies, and how this can be done at the lowest cost. Using this information, the community should decide what is the best way to establish the processing facilities, how they should be owned and managed, and who will do the work

If the intention is to increase the levels of food security in a village, the crops that are chosen for processing must be familiar to the community and the processed products should already form part of people's normal diet. The processing methods should cost as little as possible and be effective in preserving the foods for the required period. The choices that need to be considered and the factors that influence these choices are summarized in Table 3.

Table 3. Factors influencing choice when processing for food security/improved nutrition

Choices	*Factors that influence decisions*
What foods to process?	Nature and extent of nutritional deficiencies, or causes of food insecurity, types of locally avail able foods.
What type of process to use?	The local acceptability of the processed food and required storage life, types of equipment required, availability of resources to establish production and maintain/repair of equipment, protection required by food from storage structures or packaging and local availability of suitable materials.
How to make sufficient amounts of processed food with the required quality?	Extent of knowledge and skills to process the food safely in sufficient quantities and make products of acceptable quality; amount of training and technical assistance required.
Who owns, manages and operates the facilities?	Degree of cohesion and cooperation between families in the community, and willingness to invest time and resources in community ventures or individual household production.

Processing for Sale

If the purpose of introducing food processing is to generate income for families or a community, a number of decisions need to be taken before production starts. The first set of decisions concern the type of product to make. In general, the products that are best suited for village processing have a relatively high value and low volume. Similarly, high-value products that are made from cheap raw materials offer the best opportunities to add value and generate good profits. Typically fresh fruits, vegetables, cereals and root crops cost little during the harvest season, but can be processed into a range of high-value juices, pickles, baked goods, snack foods, dried foods, etc.

This added value means that for a given level of income, a relatively small amount of food must be processed, and hence the size of equipment and the required investment can be kept at affordable levels.

Table 4. Factors that influence choices when processing foods for sale

hoices	*Factors that influence decisions*
'hat foods to 'ocess?	Types, amounts, cost and quality of locally grown foods, suitability of varieties for processing, estimated size of the current and future demand for the product.
'hat type of 'ocess to use?	Extent of knowledge and skills to operate a process, resources to establish production and maintain/repair equipment, local availability, sources and costs of equipment, ingredients, packaging and distribution vehicles, requirements of the process for energy and clean water supplies, any waste disposal or air/water pollution issues.
'hat facilities are ›eded and where ıould they be cated?	Availability of a suitable production site near to raw material supply with access road and essential services (power, water, fuel, etc.); and production capacity required.
'ho owns, anages and ›erates the cilities?	Degree of cohesion and cooperation between families in the community, willingness to invest time and resources in community ventures or individual family based production.
'hat scale of 'oduction to ıoose?	Size of demand for products and share of market (from market investigation), knowledge and skills to plan production and produce sufficient quantities of food, numbers of trained production and administration staff and skills required, availability and cost of processing equipment with the required throughput, and amount of technical assistance required.
ow to make food ith the required ıality?	Extent of knowledge and skills to process the food safely to produce the quality required by customers; number of staff trained in quality assurance and, amount of training and technical assistance required.
'hat marketing ıd selling chniques to use?	Types of consumers, choice of advertising and promotion methods, distribution methods and sales outlets to be used, the main competitors and their marketing and selling techniques.
ow much finance needed?	Total investment costs, sources of finance, production costs, expected income, cash flow and profitability.

Product Selection

The two most important considerations are as follows.

1. *Supply side*. The potential types of crops or animals that can be grown in an area, the likely amounts that can be produced, the cost of production, the quality of the raw materials in relation to those produced in other areas and their suitability for processing.
2. *Demand side*. The nature and size of the demand that exists (or could exist in the future) for products that can be made by processing these crops or animals, and the number of competitors.

Communities, planners and policymakers need to balance supply and demand factors when making decisions on what type of processing to introduce.

Access to information is a key factor in deciding which products to make and which markets to exploit. Most rural communities are isolated and have little access to information on the type and size of demand for products, quality requirements of consumers or the trading conditions imposed by retailers and other buyers. An important role of government extension services and international development organizations is to find and distribute this type of information to rural producers.

It is important to emphasize that people who have little experience of food processing should select products that have a low risk of causing food poisoning. Acidic foods (such as yoghurt, pickles, fruit juices, jams, etc.) and most types of dried foods are considered safer than low-acid foods such as meats, milk, fish and vegetables. These latter foods are much more likely to cause food-borne illness if processing conditions are incorrect, or if there is poor hygiene by food handlers. Similarly, some types of process require higher levels of skill and expertise, or are more expensive to set up and operate than others. These factors should be taken into account when selecting a rural processing operation. The types of products that are usually suitable for village production are shown in Table 5.

Flour milling, cooking oil extraction, cassava processing and crop drying are also suitable and are described in the accompaning booklet in this serie *High hopes for post harvest.*

Each of these products:

— is relatively safe from causing food-borne illness;

— requires equipment that can often be made and repaired locally;
— is not technologically complex to process;
— often has a high demand and a high added-value;
— does not have sophisticated packaging requirements.

Table 5. Products that are likely to be suitable for village processing

Products	*Examples*
Bakery products	Cakes, breads, biscuits, buns, etc.
Beverages	Beers, wines, juices, squashes & cordials
Dried or smoked foods	Fruits, root crops, vegetables, meat, fish, etc.
Preserves	Jams, pastes, pickles, chutneys, sauces, etc.
Confectionery products	
Snack foods	
Yoghurt	

As in all small business development, once the type of product has been decided, it is necessary to examine all of the factors that influence the likely success of the proposed enterprise, which is known as conducting a feasibility study.

Problems to the Growth of Small-scale food Processing

There are multiple constraints on small-scale food processors, but the three most important areas are:

1) raw material supplies and planning production;
2) packaging supplies; and,
3) distribution and sales.

Raw Material Supplies and Planning Production

There are special problems that make food processing different to most other types of manufacturing business. For example, the composition and yield of crops and animals vary according to the variety, the climatic and soil conditions, and the actions of weather, pests and diseases. Many raw materials spoil rapidly after harvest or slaughter unless they are quickly processed (Table 6). Some are also highly seasonal, and can only be processed for part of the year. Each of these factors can cause unpredictable supplies and large variations in the quality and cost of raw materials. This in turn makes financial planning and production planning more difficult.

Table 6. Spoilage of foods food

Type of food	*Spoilage rate and storage life*
Animal products: meat, fish, milk.	Very rapid: a few hours.
Leafy vegetables: lettuce, some types of herbs	Rapid: 24 to 48 hours.
Soft fruit: raspberries, strawberries	Rapid: 24 to 48 hours.
Hard fruit: apples, citrus, pineapples, bananas	Slow: days or weeks.
Roots and tubers: yams, potatoes, cassava	Slow: days or weeks.
Seeds: spices, grain, nuts, oilseed	Very slow: months or years if dry.

Processors need to properly plan production to avoid delays in processing, which would result in spoilage of the raw material. Many processors lack the skills and knowledge to ensure that all inputs required for production are in place each day. Lack of an ingredient, staff shortages or machinery breakdowns due to lack of maintenance are each causes of delays and lost production, which can lead to failure of the business. In some areas, shortages of fuel are a major constraint and, in others, there may be seasonal shortages of water that restrict processing activities. Some processed foods also have a seasonal demand (for example, foods used in festivals and ceremonies, or those that are consumed at specific times of the year) and this further complicates the planning and operation of a food processing business.

Packaging

Packaging is used to control the shelf life of some foods, but in most developing countries there are serious shortages of plastic films, pots, jars, bottles, cartons, etc., which make packaging one of the most important constraints on small-scale food processors. Traditional packages including leaves, clay pots, etc., do not perform technically as well as plastics, metal or glass containers and may be perceived by customers as less attractive or inferior. For marketing purposes, smallscale processors also need to use attractive packaging to compete with larger producers or imported products.

Distribution and Sales

After processing, the shelf life of processed foods can vary from a few days to several months or years. This, together with the quality of feeder roads and the availability and cost of transport, determines the geographical area in which a processed food can be distributed. Products that require chilled or frozen transport and storage facilities are not suitable for village processing in countries where a cold distribution chain is difficult to set up and manage. Problems may also arise because processors have little influence over the activities of retailers and other buyers. Products may be poorly displayed or left on display after their "sell-by" date. They may be incorrectly stored, which leads to accelerated spoilage, or buyers may not pay for foods for many weeks, which causes cash flow problems for the processor. The processors therefore need to pay great attention to who distributes and sells their products, and influence them to protect and promote the products.

Other constraints on small-scale food processors can be grouped into those that are intrinsic or within the control of a business, including:

— lack of capital for expansion, inadequate management of finances;
— little market investigation or understanding of marketing concepts;
— under-developed entrepreneurial characteristics, selling or negotiating skills;
— lack of networks with suppliers and retailers;
— poor understanding of quality requirements by consumers, inadequate hygiene and quality assurance;
— poor understanding of opportunities for product diversification.
— Those that are extrinsic, or beyond the control of the business include:
— little information on alternative technologies, lack of equipment production by local metal workshops;
— insufficient support from research institutions and inappropriate training courses at teaching institutions;
— little influence over services provided by support agencies or government field staff;
— high cost of finance for small loans, land titles required for collateral on a loan;
— poor coordination between government institutions and non-governmental organizations and conflicting government policies (e.g. taxation vs promotion of small businesses).

In contrast to larger, formal companies and international conglomerates, most small-scale food processing enterprises have little political influence and do not receive government support (e.g. subsidies, foreign exchange allowances, price guarantees or access to specialist advice from government institutions). Some join associations to increase their economic or bargaining powers. Others seek external technical or financial assistance from development agencies, trade associations or government institutions to guide them to solutions for their individual problems. In South Asia and especially in India, small-scale food enterprises have been actively promoted by the central government for many years and there are many hundreds of thousands of successful rural enterprises.

Market Investigations

The demand for processed foods depends largely on the economic situation in a particular country or region. Where economic activity is growing, there is sufficient disposable income and growing urban populations, who form the main markets for commercially processed foods. Examples include the rapidly expanding economies in South Asia, the "Pacific Rim" and some countries in Latin America and Africa, where there has been a corresponding upsurge in small-scale food processing enterprises.

In other areas, low incomes and conservative eating habits create situations in which there is little demand for processed products beyond the usual staples, and in these places the introduction of diversified food processing is more difficult. Similarly, in countries that have opened their markets as part of structural adjustment programmes, the competition from imported processed foods that can be produced more cheaply because of economies of scale or subsidies, may overwhelm small-scale local processors and cause them to fail.

When processing foods for sale, one of the most important activities is to assess the demand for products and the proportion of that demand could realistically be met by a processor who is in competition with other producers. This not only guides the processor to which products to make, but also determines the size of the production facility and the level of investment required. The market investigation should include identification of the likely customers and consumers, and the types of promotion; marketing and selling techniques that are likely to be most effective for them. If a market assessment is not done, and re-done on a regular basis, a business is likely to eventually fail

Marketing Plans

Consumers' perceptions are not just about price and quality, but may also include status, enjoyment, attractiveness, convenience, health or nutrition. When this information is put together, a *marketing mix* can be used to summarize the factors that are special for a particular product. Using the information from a market assessment, processors can identify groups of people who will provide the greatest sales, then refine their product to meet the needs of customers and consumers, and develop a strategy to market their products to these particular people. This involves developing an attractive package, negotiating with retailers, distributors or other buyers, and designing and distributing promotional materials.

For rural processors, there are five types of geographical markets that they can consider supplying.

1. *The home community*. The nearest market is customers who are neighbours and other members of the village. Processors can sell foods that other families cannot make because they require special skills and/ or equipment, or because of lack of time or other reasons. The main outlets are sales from the home, sales in daily village markets or from local shops or kiosks.
2. *Local villages*. In many rural areas, people from neighbouring villagers meet in weekly markets that are held in one of the larger communities. Processors can rent a stall in the marketplace, set up roadside stalls for people who are travelling to the market, or rent a shop in the neighbouring village. Similarly, in some countries farmers transport products to rural trading centres, where they are sold to traders or middlemen. Sales outside a village are more important than the sale of foods to neighbours, because for economic development "new" money must be brought into a village, rather than recirculating existing money within a community.
3. *Rural towns*. Larger daily markets in rural towns attract people from a wider area and have a larger range of income groups who are potential customers at retail shops. There may also be sales to small hotels, bus stations, cafes, etc.
4. *Large towns/capital city*. These have more limited opportunities for inhabitants to cultivate their own foods, higher disposable incomes and therefore large numbers of potential consumers. Urban residents may

also be more willing to try out new or different foods. In some cities, expatriate and tourist populations also increase the demand for a wider range of processed foods. Sales can be made by processors from roadside stalls or rented stalls in daily markets or via retail shops, supermarkets, or wholesale traders/middlemen. There are also opportunities to sell processed food to hotels, institutions and other food processors.

5. Export markets. Some types of products that can be produced by village processors are suitable for export. However, the administrative and legal complexities of exporting require assistance, either from local agents, from development agencies or by creating well-organised producer associations. Large numbers of rural communities have benefited from sales of processed foods to fair trade organizations, which buy products such as dried fruits, nuts, cooking oils, shea butter and jams for sale in Europe and other industrialized regions. They assist communities to establish village development programmes and use the income from sales to fund improvements to community facilities.

Appropriate Equipment and Methods

There are much longer time intervals between the different production stages in agriculture when compared to food processing. Several months may elapse between ploughing, weeding and harvesting, and the size of equipment for the different tasks does not therefore need to be matched. In food processing, the stages in a process are often separated by a few minutes or hours, and all equipment must have a similar throughput to avoid delays caused by one piece of equipment being too small. It is therefore necessary to take account of the whole process when selecting equipment. Most are simple techniques, which require only household equipment such as knives and pans at micro-scale of production. Specialized equipment necessary for higher production levels is identified.

Bakery Products

There is a wide range of bakery products that can be made at a small-scale including biscuits, cookies, leavened or unleavened breads, cakes, flans, pastries, pies, pizzas, samosas and scones. Each can be made with different shapes, sizes, flavours, etc., and the range of potential products is therefore extremely large. Production of selected products is described in outline below.

Bakery products rarely cause food poisoning because the heat during baking reduces the numbers of micro-organisms to safe levels. However, products such as pies that contain meat, fish or vegetables, or cakes that contain cream have a greater risk of causing food poisoning. Pies should be stored in a refrigerator or in a hot display cabinet. Products that contain cream should be stored in a refrigerator. Careful food handling and thorough cleaning of equipment are essential to produce these products safely. Most products from small-scale bakeries are not packaged, except in simple polythene or paper bags to protect them from dust and insects. However, biscuits require more sophisticated packaging for a longer shelf life, and are packed in cartons covered with a moisture-proof and airtight film such as cellulose, polyester or polypropylene, or stored in airtight tins or jars.

The main item of equipment in a bakery is the oven, and this can be built by a local bricklayer. Ovens can be either heated internally or have a separate firebox. Internally heated ovens are simple and relatively cheap to construct, but the food can be contaminated by smoke and ash from the burning fuel. Fuel is burned on the hearth for a few hours and the embers are raked out before the dough is baked. A new fire is built to reheat the oven after a few batches. An externally heated oven consists of a brick or earth structure that contains a steel baking chamber (e.g. an oil drum) over the firebox. This design provides for continuous baking without the need to re-light the fire, has good fuel economy and no contamination of products by smoke or ash. Fuelwood is cheap or free in rural areas, but in many countries deforestation has resulted in increased costs and legal restrictions on its use. Wood also produces a light fluffy ash that can easily contaminate products. Charcoal is more expensive, but it produces an intense heat with little smoke. However, restrictions on charcoal burning may also apply in areas that are suffering deforestation. Where coal is available, this is the ideal fuel for bakery ovens.

Doughs and batters need to be evenly mixed, and a mixer is an important piece of equipment in a bakery. Mixing by hand is possible at microscale, but when larger quantities of dough are required an electric mixer is necessary. A simple proving cabinet for fermented doughs can be made from a wooden framework, covered in polythene sheeting, with a gently boiling pan of water inside. Dough is placed in the cabinet on racks for the required time. Commercial bakeries also require a range of baking trays and tins and small hand-tools, such as biscuit or pastry cutters, flour sieves, glazing brushes, rolling pins and whisks.

Breads

Leavened bread is produced by mixing together "strong" wheat flour, yeast, water, fat and salt to form dough. This is then kneaded and placed in a prover to allow the yeast to expand the dough with gas. It is then divided into the correct sized pieces and proofed again to allow the dough to relax. After "knocking back" to remove large gas bubbles and a final proof to allow the yeast to gently inflate the dough, it is baked in an oven. Differences in the amounts of ingredients and the conditions used at each stage of the process give rise to the wide variety of breads that can be produced. Dough for unleavened flat breads, tortillas, etc., is prepared and rolled thinly before baking on a hotplate. Batter-based breads are made by mixing the ingredients to form a uniform batter, and then pouring a portion on to a hotplate. After a few seconds, the bread is turned over and the other side is baked.

Biscuits dough is made by mixing flour, sugar, baking powder and egg, together with any nuts, dried fruit or flavourings that are required. It is rolled out and cut to the required shapes with biscuit cutters, and decorated as required (e.g. with sugar crystals, chocolate chips, crystallized fruit, etc.) and baked until golden brown.

Pastries are made by first boiling together water, butter and salt. Flour and then egg is added, beating all the time until the paste is smooth and shiny. It is piped on to trays as "finger" shapes, or "rounds" and baked in an oven. Alternatively, pastry cases can be filled with many different types of sweet or savoury fillings.

Cakes, scones or soda bread rely on gas produced from baking powder to produce an aerated product. There are two main methods of making cakes: 1) Fat or butter is beaten with sugar to create a light foam, and any colouring or essences are added. Then egg is mixed in and flour/baking powder is carefully folded it into the batter, and 2) sugar and egg are whisked to form a batter. The flour is carefully folded into the mixture until it is smooth. In both methods, other ingredients (fruit, nuts, etc.) are then blended into the batter and it is poured into baking tins and baked in an oven. Scones are made by rubbing in margarine into the flour and baking powder to form a crumb. Sugar and dried fruit are added to form dough. It is rolled out to the required thickness and cut into round shapes using a biscuit cutter. These are baked on a hotplate until browned on each side. Soda bread is similar to bread made with yeast, but uses baking powder to aerate the dough.

Beverages

Wines and beers. Wines are produced from fruit juice or pulp by fermenting sugars into alcohol using "wine yeasts". Almost any fruit can be used, but the most popular are pineapple, papaya, grape, passionfruit, banana, melon and strawberry. Beers are produced using different varieties of yeast, by fermenting a mash of cereal grains (sorghum, maize, etc.) that have been allowed to sprout. The main equipment is narrow-necked fermentation vessels, made from food grade plastic or glass and plugged with cotton wool or fitted with an air lock. After fermenting, wine is filtered through a muslin or nylon cloth and siphoned into clean containers and allowed to clear before it is bottled. A clearing agent such as bentonite is needed to produce a crystal-clear product with some types of fruit. Beer is filtered and either sold from the container or filled into bottles.

Juices, squashes and cordials. The consumption of juices, especially pineapple, passionfruit and citrus juices is increasing in urban centres in many countries. Juice can be extracted from fruits in a number of ways, depending on the hardness of the raw material. Soft fruits, such as melon and pawpaw, can be pressed in a fruit press, pulped using a juicer attachment to a food processor, or "dissolved" in a steamer. Citrus fruits are "reamed" to extract the juice without the bitter pith. Harder fruits, such as pineapple are peeled and pulped using a liquidizer. At large-scale operations, a pulper-finisher separates skins and seeds from the pulp. When a clear juice is required, it is filtered through a fine cloth or stainless steel juice strainer. Juices are filled into bottles and pasteurized in hot water. Juice production from seasonal fruits can be spread over a large part of the year by processing a sequence of fruits, or by part-processing pulps and storing them in a preservative such as sodium metabisulphite solution. Squashes are made from fruit juice mixed with sugar syrup. Cordials are crystal-clear squashes. A sugar syrup is heated to boiling and mixed with filtered juice in a stainless steel pan, before filling into bottles. These drinks are diluted with water and used a little at a time, so they may contain a preservative (usually sodium benzoate) to prevent spoilage after opening.

Confectionery

Sweets are made by boiling sugar syrup to remove water. Awide variety of confectionery products are possible by varying the ingredients, the temperature of boiling, and the method of shaping the sweets. Fondant is made by boiling sugar syrup with glucose syrup. It is then cooled and beaten to reduce the size of the sugar crystals. Creams are fondants that are diluted with a weak sugar solution. They have a shorter shelf life because of the higher water content. Gelatine sweetsincluding gums, jellies, pastilles, and marshmallows have a spongy texture, which is set by gelatine. Toffee and caramelsare made from sugar, milk and fat. Hard-boiled sweets are made from sugar syrup with added flavourings and colourings, which is boiled to a high temperature and then cooled to form hard products. For each product, the temperature of boiling can be measured using a sugar thermometer, or a sample of product is cooled and, when cold, it is checked for the correct texture.

The boiled mass is poured onto a metal, stone, or marble table to cool the product uniformly before it is shaped by hand and cut into pieces. Sweets can also be made using moulds made from rubber, plastic, metal or wood. Starch moulds are made by making impressions in a tray of corn flour, using wooden shapes. In areas of high humidity, unpackaged sweets absorb moisture from the air and become sticky. Waxed paper, aluminium foil or plastic films are used to wrap individual sweets by hand. Sweets can also be stored in glass jars, or tins with close-fitting lids.

Dried or Smoked Foods

*Drying.*Dried foods can be "high-volume, lower-value" (e.g. staple cereals or root crops), or "low-volume, highervalue" foods (dried fruits, vegetables, herbs and spices and dried meat and fish). This second category offers better opportunities for profitable production by small-scale processors. Drying removes most of the water from foods to extend their shelf life and to increase their convenience and value. The loss of weight also makes transport cheaper and easier. Care and attention to hygiene are essential during drying, especially when processing low-acid foods such as meat or fish. The temperature during drying is not usually high enough to kill any bacteria or moulds that contaminate the food and, when the food is rehydrated, they can grow again and cause spoilage or food poisoning. Polythene film is only suitable for storing dried or smoked foods for a short time before they pick up moisture, soften and go mouldy. Polypropylene gives a longer shelf life, but it is usually more expensive and it may not be available in many countries. Sealed earthenware pots or metal tins are suitable alternatives for home storage.

Blanching prevents changes in colour, flavour and texture during storage of dried vegetables. Sliced vegetables are heated for a short time in a wire basket in hot water. Sulphur dioxide prevents browning in dried foods such as apple, apricot, pineapple, root crops and coconut. In sulphuring cut or shredded foods are exposed to burning sulphur in an enclosed cabinet. In sulphiting the food is soaked in a solution of sodium sulphite or sodium metabisulphite. dioxide. In some industrialized countries, there is increasing consumer resistance to fruits that are treated with sulphur, and if products are considered for export the local export development board or export agents should be consulted about its use. Dipping in citric acid, lemon or lime juice is also sometimes used to prevent browning of light coloured foods. Syrup pre-treatment can be used to extract some of the water from fruit pieces before drying. Fruit is boiled for a few minutes in syrup and soaked overnight. This produces a succulent, soft texture in the dried product.

Sun drying is only possible in areas where the weather allows foods to be fully dried immediately after harvest. The main problems are:

— damage by pests;

— slow drying in cloudy weather and at night;

— no protection from rain or dew, which causes mould growth;

— variable quality products due to over- or under-drying; and

— large areas of land needed for the shallow layers of food.

Solar dryers protect foods from dust and pests and they increase the rate of drying compared to sun drying. They can be constructed from locally available materials at relatively low cost and there are no fuel costs. However, the extra investment in dryer construction may not be recovered by higher income if local consumers are not willing to pay higher prices for improved quality. Highvalue products, particularly for export, may justify investment in a fuel-fired dryer. These are more complex and expensive to build or maintain, and require skilled labour for their operation. A compromise is to fit a fuel-fired heater to a solar dryer to give better control over drying conditions, to allow drying at night or in cloudy weather and produce a higher quality product.

Smoking. Two methods of smoking (cold and hot smoking) are used. Cold smoking changes the colour and flavour of the food, but it is not cooked or preserved by the process. Hot smoking cooks and dries the product as well as changing the colour and flavour. Locally-constructed smokers can be made from wood or brick. They may have an external smoke generator or the smoke can be produced in the smoking chamber. "Forced draught" smokers have a fan between the firebox and the smoking chamber or in the chimney. These give greater control over the time and temperature of smoking. With all types of smokers, the temperature and smoke density can be controlled by adjusting the air supply to the smouldering wood.

Preserves

Jams, jellies and marmalades

Jam is made from fruit pulp or juice with added sugar and sometimes added pectin or citric acid. Jellies are crystal-clear jams made from juice that is filtered through a fine cloth, instead of using fruit pulp. Marmalades are produced from clear citrus juices and have fine shreds of peel suspended in the gel. The high sugar content and acidity of all preserves prevents mould growth after opening the pack so that it can be used a little at a time. These products are made by boiling fruit pulp/juice with sugar in a stainless steel pan to concentrate the mixture. For some fruits that have little natural pectin (e.g. melon), powdered or liquid pectin is added. The correct point to stop boiling can be checked using a jam thermometer, by placing a drop of the product in cold water to see if it sets, or by using a hand-held refractometer. However, a refractometer is likely to be too expensive for many small-scale processors. Preserves are hot-filled into glass or polypropylene containers using jugs and funnels.

Pastes, purees and fruit "cheeses"

The most common types of pastes and purees are tomato and garlic, which are widely used in cooking. The high solids content and natural acidity preserve them for several weeks, provided they are properly packaged. They can be made at a small-scale by carefully boiling the pulp in a pan with constant stirring to prevent it darkening or burning. An alternative method of making tomato paste is to hang the pulp in a sterilized cotton sack and allow the watery juice to leak out. Salt is then mixed in and it is hung again until the weight falls to one third of the original. This method produces a product that has a natural flavour and uses considerably less fuel than boiling. Fruit cheeses are fruit pulps that are boiled until they set as a solid block when cooled. They can be cut into bars or cubes to eat directly or small pieces can be used in confectionery or bakery products.

Pickles and chutneys

There is a wide variety of pickled fruits and vegetables. Fermented pickles are made from vegetables such as cucumber, cabbage, olive or onion using lactic acid bacteria to ferment sugars in the food to lactic acid. Unfermented pickles are vegetables packed in jars with vinegar, salt and sometimes sugar and pasteurized in a bath of hot water. Sweet pickles are made from single fruits or mixtures of fruits and vegetables. They are preserved by the combined action of lactic acid or vinegar, sugar and in some cases added spices. Pickles may be packed in small quantities in sealed polythene pouches, or in plastic pots. Salted vegetables are made in a drum by building up alternate layers of chopped or shredded vegetable (e.g. cabbage) with layers of salt. The salt draws out water from the vegetables to form a concentrated brine, so products can be kept for several months. The salt is reduced by washing the vegetables before they are eaten. Chutneys are thick, jam-like mixtures made from a variety of fruits and vegetables, sugar, spices and sometimes vinegar. The high sugar content and acidity preserves the product. Some products are boiled which pasteurizes them, and others are allowed to ferment so that acids produced by the bacteria preserve the product. Some spices such as ginger, mustard, chilli and garlic also have a preservative effect. In the process, ingredients are boiled until thickened and packaged in a similar way to pickles.

Sauces

Tomato sauce, chilli sauce and mixed fruit sauces such as "Worcester" sauce are thick liquids that are made from pulped fruit and/or vegetables. They are boiled together with salt, sugar, vinegar and spices (e.g. cinnamon, cloves, allspice or cayenne pepper). The acidity, salt and sugar prevent spoilage after the pack is opened. On a small scale, the ingredients are heated in stainless steel pans with constant stirring to avoid burning the product, and packed in bottles.

Snack Foods

Sliced root crops (yam, cassava, sweet potato, Irish potato, etc.) and a number of starchy fruits, (jackfruit, breadfruit or banana) can be fried and eaten as snack foods. No specialized equipment is required at microscale production, but on a larger scale, a deepfat fryer and a slicing machine can be used to produce more uniform products and reduce the time and effort involved in processing. They can be stored for a short time in polythene bags, but rapidly develop a rancid flavour. Polypropylene film is needed for longer storage. Doughnuts are made by fermenting dough made from water, flour, sugar and yeast and made into balls. Ring doughnuts are made by removing the centre from the dough piece with a cutter. The dough is fried in deep oil until all sides are golden brown. Doughnuts can be coated with sugar or filled with jam or cream.Samosas are triangular packages of "filo" pastry, which are filled with vegetables or ground meat or fish. They are fried until heated to the centre and stored in a hot cabinet until consumed.

Yoghurt

Yoghurt is produced by fermenting milk using two species of bacteria that change milk sugar (lactose) into lactic acid. This forms the characteristic curd and restricts spoilage so that yoghurt is preserved for up to ten days in a refrigerator. Milk is a low acid food, which can contain food poisoning bacteria. It should be handled hygienically and kept cool until it is used. In the process milk is pasteurized and cooled to body temperature, and a starter culture of the bacteria is added. Either the starter can be bought as a powder or it can be taken from a previous batch of product. A simple incubator can be made using a block of thick polystyrene to hold the pots, which maintains the temperature of the product for several hours. Alternatively, pots are placed in an insulated box, fitted with an electric light bulb. The heat from the bulb maintains the temperature. Set yoghurt is made by filling pots with milk after adding the starter. For liquid yoghurt, the yoghurt is stirred to break the gel and then poured into pots. Chopped fruits or nuts can be added to each type of yoghurt, but care is needed to ensure that they are thoroughly cleaned to avoid contamination by moulds. Less acidic fruits such as melon or pawpaw are more successful because they do not react with the milk. Acidic fruits (lemon, lime, pineapple, etc.) may cause curdling and separation of the yoghurt.

Resources and Services Required for Successful Processing

The resources and services required for successful exploitation of markets by small-scale processors include the following.

1. A suitable processing room with reliable services, including clean water and for some products, electricity and fuel supplies.
2. Access to raw materials, ingredients and packaging.
3. Sources of affordable equipment and local servicing skills.

4. Adequate transport, good road infrastructure and feeder roads.
5. Training, skill development and networking.
6. Access to technical backup for advice on topics such as quality assurance.
7. Good linkages with suppliers and sellers

Production Facilities

The building

All food processing businesses should have a hygienic and easily cleaned building to prevent contamination of products. It should be on a fenced site to keep out animals, with short grass to trap airborne dust. The investment in construction or modification of a building should be appropriate to the size and expected profitability of the business. Within the building, foods should move between different stages in a process without the paths crossing, to reduce the risk of contaminating finished products by incoming often dirty, raw materials. There should be enough space for separate storage of ingredients, packaging materials and finished products.

Overhanging roofs keep a building cool and fibre-cement tiles provide greater insulation than iron sheets against heat from the sun. A panelled ceiling should be fitted, rather than exposed roof beams, which allow dust to accumulate and contaminate products. Beams are also paths for rodents and birds, creating contamination risks from hairs, feathers or excreta. It is important to ensure that there are no holes in the ceiling or roof, and no gaps where the roof joins the walls that would allow pests to enter. All internal walls should be plastered or rendered with concrete. They should have no cracks that could harbour dirt or insects. The lower parts of the walls are most likely to get dirty and they should either be tiled or painted with white gloss paint. Higher parts of walls can be painted with emulsion paint.

Natural daylight is the preferred lighting in processing rooms. The number and size of windows depends on how much money a processor wishes to invest, and the security risk in a particular area (windows are more expensive than walls, especially when security bars or grilles are needed). Storerooms do not need to have windows. All windows should be screened with mosquito mesh. Doors should not have gaps beneath them, which would allow pests to enter. If the doors are used regularly, thin metal chains or strips of plastic can be hung from door lintels, or mesh door screens can be fitted to deter insects and some animals.

Floors should be made of good quality concrete, without holes or cracks. The floor should be curved up to meet the wall to prevent dirt collecting in corners. Except in dry processes (e.g. milling or baking), the floor should slope to a drainage channel, fitted with metal gratings that are easily removed for cleaning. A wire mesh cover should be fitted over the drain opening to prevent rats and crawling insects from entering the room.

Services

The availability of clean water, electricity, fuels, servicing and maintenance facilities and transport varies considerably in different countries and between regions of the same country. Water is essential in all food processing, either as an ingredient or for washing equipment. An adequate supply of safe water should be available from taps in the processing room. There is unlikely to be a mains water supply in many rural areas, and water from boreholes is the preferred choice. It is likely to be less contaminated with micro-organisms than river water, but it may contain sand. River water should only be used if no other source is available. To remove sand, two high level covered storage tanks should be installed, either in the roof-space or on pillars outside the building. While one tank is being used any sand in the other tank settles out. If necessary, water should be treated to remove micro-organisms by dosing it with bleach.

Good sanitation is essential to prevent contamination of products, and equipment should be thoroughly cleaned after each day's production. All wastes should be disposed of in a way that does not damage the local environment. Solid wastes should be removed from the building at regular intervals, and either buried or turned into animal feed or compost. In processes that produce large amounts of liquid wastes, these should be carefully disposed of to prevent local pollution of streams or lakes. A soakaway should be constructed that cannot contaminate drinking water supplies. Toilets should be separated from the processing area or be in a separate building. Workers should have hand-washing facilities with soap and clean towels, and they should receive training in hygiene and safe food handling.

Electric power is needed at larger scales of processing for some products. If mains electricity is not available, the main alternative is a diesel powered generator, although the cost may be prohibitive for many processors. Support agencies could also investigate the local potential of wind- or water-powered turbines. Photovoltaic cells may be useful for lighting, for small refrigerators or for topping up batteries, but they do not supply sufficient power to run most processing equipment. Where electric power is available, sockets should be located high enough on walls to prevent them getting wet when equipment is washed down. All plugs should have appropriate fuses for the power rating of the equipment, and the mains supply should have an earth leakage trip-switch. Where lighting is needed, florescent tubes use less electricity than light bulbs.

Fuels are required for processes that involve heating (boiling, drying, smoking, etc.) and these can be an important constraint in some areas. This is particularly the case where wood is the only available fuel and supplies are reduced because of deforestation. Gas does not contaminate products with smoke, it is easily controllable, has a high heat output, and cylinders are transportable for re-filling. Where gas cylinders are not available, a biogas generator can be constructed to provide fuel, if the use of manure does not detract from its use as fertilizer and if there are sufficient quantities to produce enough gas for processing. Solar heaters may also be useful to pre-heat water.

Transport is needed for incoming raw materials, ingredients and packaging, and for outgoing products. Most small-scale processors are able to hire a pickup truck when needed, but they have little control over the conditions during transport, which may cause damage to packages or contaminate foods (e.g. by transporting them with non-food materials, live animals or contaminating them with grease or oils). Feeder roads to villages are often in poor condition, and may only be open during part of the year. Delays caused by poor quality roads add to the cost of processing and cause damage to raw materials and packages (especially glass containers). The condition of roads is an important consideration when selecting a site for processing, if it is intended to supply markets other than the local village.

Servicing and maintenance facilities

Most domestic equipment used for micro-scale processing can be maintained by careful use and regular cleaning. Specialist equipment such as presses, ovens, etc., should ideally be made by local workshops that can also maintain them, or there should be a supplier of spare parts in a local town. Local equipment manufacture may require the collaboration of food research institutes and university food technology departments to develop and test prototypes so that they meet the needs of both workshops and food processors. Where imported equipment is used, a supply of spare parts should be provided, and the business owner should be trained to correctly replace worn-out parts.

Small-scale food processing requires reliable, affordable, locally produced and locally repaired, technology of a suitable size for the people who operate it.

Training, Skill Development and Networking

One approach to improved village processing is to upgrade traditional products to improve their quality or shelf life and hence increase sales in more distant markets. Another approach is to produce non-traditional foods from existing crops or animals.

Both types of processing require support through training, advice and skill development. When successful this support can lift rural people out of poverty by increasing their skills, confidence, knowledge and resources, and by providing diversified opportunities for them to process and sell their products. However, in practice, the consequences of introducing food processing to a rural community may be difficult to predict.

Although potential adverse effects of a new technology on rural producers can to some extent be predicted and avoided by careful studies before a project is implemented, there are a large number of factors, which affect the final outcome and determine who will benefit from technological changes. There is thus a need for sensitivity and understanding by planners and field workers of the social and cultural context in which the changes are introduced.

Advisers and planners have a responsibility to carefully evaluate processing technologies to ensure that they are effective in meeting the needs of the individual communities that are being assisted. The evaluation criteria are complex and inter-related and may differ in different communities, but a summary of the main aspects is given in Table 7.

Food processing developments should build on existing skills, traditional knowledge and practices, and not cause abrupt changes to village life styles or cultures. It is important that the criteria in Table 7 are not simply used by advisers as a checklist, but should be used to assist their judgement of the needs and solutions required by each individual community. Care should also be taken to ensure that the activities can be accommodated within family routines, without excessively increasing the workload on individual family members, and most importantly that there is sufficient profit to make the venture worthwhile. Poor communities with few resources are vulnerable to risk and advisers should make every effort to minimize the risks inherent in new ventures by conducting careful feasibility studies.

Table 7. Some criteria for assessing the suitability of processing technologies for rural communities

- Conformity to local traditions, beliefs and food habits.
- Conformity with existing administrative or social structures.
- Technical effectiveness - can the process produce foods in the required amount and to the required quality standards?
- Costs of purchase and maintenance/repair of equipment and any ancillary services required.
- Cost and availability of fuels and power; alternative sources of energy (e.g. solar, biomass, wind or water power).
- Operating costs, expected sales income and overall financial profitability.
- Distribution of profits within the community.
- Effect of increase in employment or displacement of workers.
- Training and skill levels required.
- Health and safety issues.
- Environmental impact

Village scale processing can be done either by individual families or by cooperation from a number of families. If a cooperative approach is chosen, the residents of a village need to decide how they wish their community to develop, how they want to benefit from food processing, and how these developments should be managed within the community. As part of this process the benefits of joining forces with other communities, or national and international groups should also be explored.

Lack of education, rural isolation, and adverse social structures may each contribute to poorly developed production and selling skills in rural communities. The lack of direct contact with consumers means that villagers are often not aware of their changing quality requirements and, as a result, do not produce foods to the standards of hygiene or quality that are required. Training programmes that are held as part of a food processing development programme should include confidence-building techniques, financial management and marketing skills as well as the technical skills needed to produce high quality processed foods.

References

Fellows, P.J. & Axtell, B.L.A. 2001. *Opportunities in food processing - setting up and running a small food business*. CTA Publications, Wageningen, the Netherlands.

Fellows, P.J. 2002, *Promoting linkages between food producers and processors.* Booklet, FAO, Rome, Italy.

Good, A. 1997. Fruit factory in the forest, *Food Chain* 21, 4-5.

Herklots, J. 1991. Making money from honey, *Food Chain,* 3, 3-5.

Hidellage, V. 1999. Empowering small-scale cashew processors in Sri Lanka, *Food Chain,* 24, 11-15.

Jayaraj, J. 1999. Training in food processing - a sustainable approach in India, *Food Chain,* 24, 19-21.

Trager, J. 1996. *The Food Chronology,* Aurum Press, London, UK

3

Processing of Vegetables

The main objective of vegetable processing is to supply wholesome, safe, nutritious and acceptable food to consumers throughout the year. Vegetable processing projects also aim to replace imported products like squash, yams, tomato sauces, pickles, etc., besides earning foreign exchange by exporting finished or semi-processed products.

The vegetable processing activities have been set up, or have to be established in developing countries for one or other of the following reasons:

- diversification of the economy, in order to reduce present dependence on one export commodity;
- government industrialisation policy;
- reduction of imports and meeting export demands;
- stimulate agricultural production by obtaining marketable products;
- generate both rural and urban employment;
- reduce vegetable losses;
- improve farmers' nutrition by allowing them to consume their own processed vegetables during the off-season;
- generate new sources of income for farmers/artisans;
- develop new value-added products.

Practically any vegetable can be processed, but some important factors which determine whether it is worthwhile are:

a. the demand for a particular fruit or vegetable in the processed form;

b. the quality of the raw material, i.e. whether it can withstand processing;

c. regular supplies of the raw material.

For example, a particular variety of fruit which may be excellent to eat fresh is not necessarily good for processing. Processing requires frequent handling, high temperature and pressure.

Many of the ordinary table varieties of tomatoes, for instance, are not suitable for making paste or other processed products. A particular mango or pineapple may be very tasty eaten fresh, but when it goes to the processing centre it may fail to stand up to the processing requirements due to variations in its quality, size, maturity, variety and so on.

Even when a variety can be processed, it is not suitable unless large and regular supplies are made available. An important processing centre or a factory cannot be planned just to rely on seasonal gluts; although it can take care of the gluts it will not run economically unless regular supplies are guaranteed.

To operate a vegetable processing centre efficiently it is of utmost importance to pre-organise growth, collection and transport of suitable raw material, either on the nucleus farm basis or using outgrowers.

Processing Planning

The secret of a well planned vegetable processing centre is that it must be designed to operate for as many months of the year as possible. This means the facilities, the buildings, the material handling and the equipment itself must be inter-linked and coordinated properly to allow as many products as possible to be handled at the same time, and yet the equipment must be versatile enough to be able to handle many products without major alterations.

A typical processing centre or factory should process four or five types of fruits harvested at different times of the year and two or three vegetables. This processing unit must also be capable of handling dried/dehydrated finished products, juices, pickles, tomato juice, ketchup and paste, jams, jellies and marmalades, semi-processed fruit products.

Advanced planning is necessary to process a large range of products in varied weather and temperature conditions, each requiring a special set of manufacturing and packaging formulae. The end result of the efforts should be a well-managed processing unit with lower initial investment.A unit which is sensibly laid out and where one requirement co-relates to another, with a sound costing analysis, leads to an integrated operation.

Instead of over-sophisticated machinery, a sensible simple processing unit may be required when planned production is not very large and is geared mainly to meet the demand of the domestic market.

General Properties of Vegetables

Vegetables have many similarities with respect to their compositions, methods of cultivation and harvesting, storage properties and processing. In fact, many vegetables may be considered fruit in the true botanical sense. Botanically, fruits are those portions of the plant which house seeds. Therefore such items as tomatoes, cucumbers, eggplant, peppers, and others would be classified as fruits on this basis.

However, the important distinction between vegetables has come to be made on an usage basis. Those plant items that are generally eaten with the main course of a meal are considered to be vegetables. Those that are commonly eaten as dessert are considered fruits.

Vegetables are derived from various parts of plants and it is sometimes useful to associate different vegetables with the parts of the plant they represent since this provides clues to some of the characteristics we may expect in these items.

Fruit as a dessert item, is the mature ovaries of plants with their seeds. The edible portion of most fruit is the fleshy part of the pericarp or vessel surrounding the seeds. Fruit in general is acidic and sugary. They commonly are grouped into several major divisions, depending principally upon botanical structure, chemical composition and climatic requirements.

Berries are fruit which are generally small and quite fragile. Grapes are also physically fragile and grow in clusters. Melons, on the other hand, are large and have a tough outer rind. Drupes (stone fruit) contain single pits and include such items as apricots, cherries, peaches and plums. Pomes contain many pits, and are represented by apples, quince and pears.

Citrus fruit like oranges, grapefruit and lemons are high in citric acid. Tropical and subtropical fruits include bananas, dates, figs, pineapples, mangoes, and others which require warm climates, but exclude the separate group of citrus fruits.

Compositions of vegetables and fruit not only vary for a given kind in according to botanical variety, cultivation practices, and weather, but change with the degree of maturity prior to harvest, and the condition of ripeness, which is progressive after harvest and is further influenced by storage conditions. Nevertheless, some generalisations can be made.

Most fresh vegetables and fruit are high in water content, low in protein, and low in fat. In these cases water contents will generally be greater than 70% and frequently greater than 85%.

Commonly protein content will not be greater than 3.5% or fat content greater than 0.5 %. Exceptions exist in the case of dates and raisins which are substantially lower in moisture but cannot be considered fresh in the same sense as other fruit. Legumes such as peas and certain beans are higher in protein; a few vegetables such as sweet corn which are slightly higher in fat and avocados which are substantially higher in fat.

Vegetables and fruit are important sources of both digestible and indigestible carbohydrates. The digestible carbohydrates are present largely in the form of sugars and starches while indigestible cellulose provides roughage which is important to normal digestion.

Vegetables are also important sources of minerals and certain vitamins, especially vitamins A and C. The precursors of vitamin A, including beta-carotene and certain other carotenoids, are to be found particularly in the yellow-orange vegetables and in the green leafy vegetables.

Citrus fruit are excellent sources of vitamin C, as are green leafy vegetables and tomatoes. Potatoes also provide an important source of vitamin C for the diets of many countries. This is not so much due to the level of vitamin C in potatoes which is not especially high but rather to the large quantities of potatoes consumed.

Chemical Composition

Water

Vegetal cells contain important quantities of water. Water plays a vital role in the evolution and reproduction cycle and in physiological processes. It has effects on the storage period length and on the consumption of tissue reserve substances.

In vegetal cells, water is present in following forms:

— bound water or dilution water which is present in the cell and forms true solutions with mineral or organic substances;
— colloidal bound water which is present in the membrane, cytoplasm and nucleus and acts as a swelling agent for these colloidal structure substances; it is very difficult to remove during drying/dehydration processes;
— constitution water, directly bound on the chemical component molecules and which is also removed with difficulty.

Vegetables contain generally 90-96% water while for fruit normal water content is between 80 and 90%.

Mineral substances

Mineral substances are present as salts of organic or inorganic acids or as complex organic combinations (chlorophyll, lecithin, etc.); they are in many cases dissolved in cellular juice.

Vegetables are more rich in mineral substances as compared with fruits. The mineral substance content is normally between 0.60 and 1.80% and more than 60 elements are present; the major elements are: K, Na, Ca, Mg, Fe, Mn, Al, P. Cl, S.

Among the vegetables which are especially rich in mineral substances are: spinach, carrots, cabbage and tomatoes. Mineral rich fruit includes: strawberries, cherries, peaches and raspberries. Important quantities of potassium (K) and absence of sodium chloride (NaCl) give a high dietetic value to fruit and to their processed products. Phosphorus is supplied mainly by vegetables.

Vegetables usually contain more calcium than fruit; green beans, cabbage, onions and beans contain more than 0.1% calcium. The calcium/phosphorus or Ca/P ratio is essential for calcium fixation in the human body; this value is considered normal at 0.7 for adults and at 1.0 for children. Some fruit are important for their Ca/P ratio above 1.0: pears, lemons, oranges and some temperate climate mountain fruits and wild berries.

Even if its content in the human body is very low, iron (Fe) has an important role as a constituent of haemoglobin. Main iron sources are apples and spinach.

Carbohydrates

Carbohydrates are the main component of vegetables and represent more than 90% of their dry matter. From an energy point of view carbohydrates represent the most valuable of the food components; daily adult intake should contain about 500 g carbohydrates.

Carbohydrates play a major role in biological systems and in foods. They are produced by the process of photosynthesis in green plants. They may serve as structural components as in the case of cellulose; they may be stored as energy reserves as in the case of starch in plants; they may function as essential components of nucleic acids as in the case of ribose; and as components of vitamins such as ribose and riboflavin.

Carbohydrates can be oxidised to furnish energy, and glucose in the blood is a ready source of energy for the human body. Fermentation of carbohydrates by yeast and other microorganisms can yield carbon dioxide, alcohol, organic acids and other compounds.

Fats

Generally vegetables contain very low level of fats, below 0.5%. However, significant quantities are found in nuts (55%), apricot kernel (40%), grapes seeds (16%), apple seeds (20%) and tomato seeds (18%).

Organic acids

Fruit contains natural acids, such as citric acid in oranges and lemons, malic acid of apples, and tartaric acid of grapes. These acids give the fruits tartness and slow down bacterial spoilage.

We deliberately ferment some foods with desirable bacteria to produce acids and this give the food flavour and keeping quality. Examples are fermentation of cabbage to produce lactic acid and yield sauerkraut and fermentation of apple juice to produce first alcohol and then acetic acid to obtain vinegar.

Nitrogen-containing substances

These substances are found in plants as different combinations: proteins, amino acids, amides, amines, nitrates, etc. Vegetables contain between 1.0 and 5.5 % while in fruit nitrogen-containing substances are less than 1% in most cases. Among nitrogen containing substances the most important are proteins; they have a colloidal structure and, by heating, their water solution above 50°C an one-way reaction makes them insoluble.

Vitamins

Vitamins are defined as organic materials which must be supplied to the human body in small amounts apart from the essential amino-acids or fatty acids.

Vitamins function as enzyme systems which facilitate the metabolism of proteins, carbohydrates and fats but there is growing evidence that their roles in maintaining health may extend yet further.

The vitamins are conveniently divided into two major groups, those that are fat-soluble and those that are water-soluble. Fat-soluble vitamins are A, D, E and K. Their absorption by the body depends upon the normal absorption of fat from the diet. Water-soluble vitamins include vitamin C and several members of the vitamin B complex.

Enzymes

Enzymes are biological catalysts that promote most of the biochemical reactions which occur in vegetable cells. Enzymes have an optimal temperature - around +50°C where their activity is at maximum. Heating beyond this optimal temperature deactivates the enzyme. Activity of each enzyme is also characterised by an optimal pH.

In vegetable storage and processing the most important roles are played by the enzymes classes of hydrolases (lipase, invertase, tannase, chlorophylase, amylase, cellulase) and oxidoreductases (peroxidase, tyrosinase, catalase, ascorbinase, polyphenoloxidase).

Structural Features

The structural unit of the edible portion of most fruits and vegetables is the parenchyma cell. While parenchyma cells of different vegetables differ somewhat in gross size and appearance, all have essentially the same fundamental structure.

Parenchyma cells of plants differ from animal cells in that the actively metabolising protoplast portion of plant cells represents only a small fraction, of the order of five per cent, of the total cell volume. This protoplast is film-like and is pressed against the cell wall by the large water-filled central vacuole.

The protoplast has inner and outer semi-permeable membrane layers; the cytoplasm and its nucleus are held between them. The cytoplasm contains various inclusions, among them starch granules and plastics such as the chloroplasts and other pigment-containing chromoplasts. The cell wall, cellulose in nature, contributes rigidity to the parenchyma cell and limits the outer protoplasmic membrane. It is also the structure against which other parenchyma cells are cemented to form extensive three-dimensional tissue masses.

The layer between cell walls of adjacent parenchyma cells is referred to as the middle lamella, and is composed largely of pectic and polysaccharide cement-like materials. Air spaces also exist, especially at the angles formed where several cells come together.

The relationships between these structures and their chemical compositions are further outlined below. The parenchyma cells will vary in size among plants but are quite large when compared to bacterial or yeast cells. The larger parenchyma cells may have volumes many thousand times greater than a typical bacterial cell.

There are additional types of cells other than parenchyma cells that make up the familiar structures of vegetables. These include various types of conducting cells which are tube-like and distribute water and salts throughout the plant. Such cells produce fibrous structures toughened by the presence of cellulose and the woodlike substance lignin. Cellulose, lignin, and pectic substances also occur in specialised supporting cells which increase in importance as plants become older.

An important structural feature of all plants, including vegetables is protective tissue. This can take many forms but usually is made up of specialised parenchyma cells that are pressed compactly together to form a skin, peel or rind. Surface cells of these protective structures on leaves, stems or fruit secrete waxy cutin and form a water impermeable cuticle. These surface tissues, especially on leaves and young stems will also contain numerous valve-like cellular structures, the stomata, through which moisture and gases can pass.

Deterioration Factors and Their Control

A summary of overall deterioration reactions in fruits and vegetables is presented below.

Enzymic Changes

Enzymes which are endogenous to plant tissues can have undesirable or desirable consequences. Examples involving endogenous enzymes include

a) the post-harvest senescence and spoilage of vegetables; b) oxidation of phenolic substances in plant tissues by phenolase (leading to browning); c) sugar - starch conversion in plant tissues by amylases; d) post-harvest demethylation of pectic substances in plant tissues (leading to softening of plant tissues during ripening, and firming of plant tissues during processing).

The major factors useful in controlling enzyme activity are: temperature, water activity, pH, chemicals which can inhibit enzyme action, alteration of substrates, alteration of products and pre-processing control.

Chemical changes

The two major chemical changes which occur during the processing and storage of foods and lead to a deterioration in sensory quality are lipid oxidation and non-enzymatic browning. Chemical reactions are also responsible for changes in the colour and flavour of foods during processing and storage.

Colour changes

Chlorophylls. Almost any type of food processing or storage causes some deterioration of the chlorophyll pigments. Phenophytinisation (with consequent formation of a dull olivebrown phenophytin) is the major change; this reaction is accelerated by heat and is acid catalysed.Other reactions are also possible. For example, dehydrated products such as green peas and beans packed in clear glass containers undergo photo-oxidation and loss of desirable colour.

Anthocyanins. These are a group of more than 150 reddish water-soluble pigments that are very widespread in the plant kingdom. The rate of anthocyanin destruction is pH dependent, being greater at higher pH values. Of interest from a packaging point of view is the ability of some anthocyanins to form complexes with metals such as Al, Fe, Cu and Sn.

Carotenoids. The carotenoids are a group of mainly lipid soluble compounds responsible for many of the yellow and red colours of plant and animal products. The main cause of carotenoid degradation in foods is oxidation. The mechanism of oxidation in processed foods is complex and depends on many factors. The pigments may auto-oxidise by reaction with atmospheric oxygen at rates dependent on light, heat and the presence of pro- and antioxidants.

Flavour changes

In vegetables, enzymically generated compounds derived from long-chain fatty acids play an extremely important role in the formation of characteristic flavours. In addition, these types of reactions can lead to significant off-flavours. Enzyme-induced oxidative breakdown of unsaturated fatty acids occurs extensively in plant tissues and this yield characteristic aromas associated with some ripening fruits and disrupted tissues.

Nutritional quality

The four major factors which affect nutrient degradation and can be controlled to varying extents by packaging are light, oxygen concentration, temperature and water activity. However, because of the diverse nature of the various nutrients as well as the chemical heterogeneity within each class of compounds and the complex interactions of the above variables, generalizations about nutrient degradation in foods will inevitably be broad ones.

Physical changes

One major undesirable physical change in food powders is the absorption of moisture as a consequence of an inadequate barrier provided by the package; this results in caking. It can occur either as a result of a poor selection of packaging material in the first place, or failure of the package integrity during storage. In general, moisture absorption is associated with increased cohesiveness.

Biological changes

Microbiological

Micro-organisms can make both desirable and undesirable changes to the quality of foods depending on whether or not they are introduced as an essential part of the food preservation process or arise unintentionally and subsequently grow to produce food spoilage.The two major groups of micro-organisms found in foods are bacteria and fungi, the latter consisting of yeasts and moulds. Bacteria are generally the fastest growing, so that in conditions favourable to both, bacteria will usually outgrow fungi.

Foods are frequently classified on the basis of their stability as non-perishable, semi-perishable and perishable. For example, hermetically sealed and heat processed (e.g. canned) foods are generally regarded as non-perishable. However, they may become perishable under certain circumstances when an opportunity for recontamination is afforded following processing.

Such an opportunity may arise if the can seams are faulty, or if there is excessive corrosion resulting in internal gas formation and eventual bursting of the can. Spoilage may also take place when the canned food is stored at unusually high temperatures: thermophilic spore-forming bacteria may multiply, causing undesirable changes such as flat sour spoilage.

Macrobiological

Insect Pests

Warm humid environments promote insect growth, although most insects will not breed if the temperature exceeds about 35 C° or falls below 10 C°. Also many insects cannot reproduce satisfactorily unless the moisture content of their food is greater than about 11%.

The main categories of foods subject to pest attack are cereal grains and products derived from cereal grains, other seeds used as food (especially legumes), dairy products such as cheese and milk powders, dried fruits, dried and smoked meats and nuts.

As well as their possible health significance, the presence of insects and insect excrete in packaged foods may render products unsaleable, causing considerable economic loss, as well as reduction in nutritional quality, production of off-flavours and acceleration of decay processes due to creation of higher temperatures and moisture levels.

Rodents

Rats and mice carry disease-producing organisms on their feet and/or in their intestinal tracts and are known to harbour salmonella of serotypes frequently associated with food-borne infections in humans. In addition to the public health consequences of rodent populations in close proximity to humans, these animals also compete intensively with humans for food.

Methods of Reducing Deterioration

A knowledge of deterioration factors and the way they act, including the rates of deterioration to a specific category of food, means that it is possible to list the ways of lowering or stopping the action and obtaining vegetable preservation.

In order to maintain their nutritional value and organoleptic properties and because of technical-economical considerations, not all the identified means against deterioration actually have practical applications for vegetable preservation.

Technical Methods of Reducing Food Deterioration

These technical means can be summarised as follows:

Physical	*Heating*
	Cooling
	Lowering of water content
	Drying/dehydration.
	Concentration
	Sterilising filtration
	Irradiation
	Other physical means (high pressure, vacuum, inert gases)
Chemical	Salting
	Smoking
	Sugar addition
	Artificial acidification
	Ethyl alcohol addition
	Antiseptic substance action
Biochemical	Lactic fermentation (natural acidification)
	Alcoholic fermentation

Harvesting and Pre-processing

When vegetables are maturing in the field they are changing from day to day. There is a time when the vegetable will be at peak quality from the stand-point of colour, texture and flavour. This peak quality is quick in passing and may last only a day. Harvesting and processing of several vegetables, including tomatoes, corn and peas are rigidly scheduled to capture this peak quality.

After the vegetable is harvested it may quickly pass beyond the peak quality condition. This is independent of microbiological spoilage; these main deteriorations are related to:

— loss of sugars due to their consumption during respiration or their conversion to starch; losses are slower under refrigeration but there is still a great change in vegetable sweetness and freshness of flavour within 2 or 3 days;

— production of heat when large stockpiles of vegetables are transported or held prior to processing. At room temperature some vegetables will liberate heat at a rate of 127,000 kJ/ton/day; this is enough for each ton of vegetables to melt 363 kg of ice per day. Since the heat further deteriorates the vegetables and speeds microorganisms growth, the harvested vegetables must be cooled if not processed immediately. But cooling only slows down the rate of deterioration, it does not prevent it, and vegetables differ in their resistance to cold storage. Each vegetable has its optimum cold storage temperature which may be between about 0-100 C (32-500 F).

— the continual loss of water by harvested vegetables due to transpiration, respiration and physical drying of cut surfaces results in wilting of leafy vegetables, loss of plumpness of fleshy vegetables and loss of weight of both.

Moisture loss cannot be completely and effectively prevented by hermetic packaging. This was tried with plastic bags for fresh vegetables in supermarkets but the bags became moisture fogged, and deterioration of certain vegetables was accelerated because of buildup of CO_2 and decrease of oxygen in the package. It therefore is common to perforate such bags to prevent these defects as well as to minimise high humidity in the package which would encourage microbial growth.

Shippers of fresh vegetables and vegetable processors, whether they can, freeze, dehydrate, or manufacture soups or ketchup, appreciate the instability and perishability of vegetables and so do everything they can to minimise delays in processing of the fresh product. In many processing plants it is common practice to process vegetables immediately as they are received from the field.

To ensure a steady supply of top quality produce during the harvesting period the large food processors will employ trained field men; they will advise on growing practices and on spacing of plantings so that vegetables will mature and can be harvested in rhythm with the processing plant capabilities. This minimises stockpiling and need for storage.

Cooling of vegetables in the field is common practice in some areas. Liquid nitrogen-cooled trucks may next provide transportation of fresh produce to the processing plant or directly to market. Upon arrival of vegetables at the processing centre the usual operations of cleaning, grading, peeling, cutting and the like are performed using a moderate amount of equipment but a good deal of hand labour also still remains.

Reception

This covers qualitative and quantitative control of delivered vegetables. The organoleptic control and the evaluation of the sanitary state, even if they are very important steps in vegetables' characteristics assessment, cannot establish their technological value. On the other hand, laboratory controls do not precisely establish their technological properties because of the difficulty in putting into showing some deterioration when using rapid control methods.

One correct method of vegetable quality appraisal is their overall evaluation based on the whole complex of data that can be obtained by combining an extensive organoleptic evaluation with simple analysis that can be performed rapidly in plant laboratory. These analysis can be:

- refractometric extract (tomatoes, fruit, etc.);
- specific weight (potatoes, peas, etc.);
- consistency (measured with tenderometers, penetrometers, etc.);
- boiling tests, etc.

Temporary Storage

This step should be as short as possible and better completely eliminated. Vegetables can be stored in:

- simple stores, without artificial cooling;
- in refrigerated stores; or, in some cases,
- in silos (potatoes, etc.).

Simple stores should be covered, fairly cool, dry and well ventilated but without forced air circulation which can induce significant losses in weight through intensive water evaporation; air relative humidity should be at about 70-80%. Refrigerated storage is always preferable and in all cases a processing centre needs a cold room for this purpose.

Washing Equipment

Washing is used not only to remove field soil and surface microorganisms but also to remove fungicides, insecticides and other pesticides, since there are laws specifying maximum levels of these materials that may be retained on the vegetable; and in most cases the allowable residual level is virtually zero. Washing water contains detergents or other sanitisers that can essentially completely remove these residues.

The washing equipment, like all equipment subsequently used, will depend upon the size, shape and fragility of the particular kind of vegetable:

— flotation cleaner for peas and other small vegetables;

— rotary washer in which vegetables are tumbled while they are sprayed with jets of water; this type of washer should not be used to clean fragile vegetables;

Sorting

This step covers two separate operations:

— removal of non-standard vegetables (and fruit) and possible foreign bodies remaining after washing;

— quality grading based on variety, dimensional, organoleptical and maturity stage criterion.

Skin Removal

Some vegetables require skin removal. This can be done in various ways.

Mechanical

This type of operation is performed with various types of equipment which depend upon the result expected and the characteristics of the fruit and vegetables, for example:

— a machine with abrasion device (potatoes, root vegetables);

— equipment with knives (apples, pears, potatoes, etc.);

— equipment with rotating sieve drums (root vegetables). Sometimes this operation is simultaneous with washing (potatoes, carrots, etc.) or preceded by blanching (carrots).

Chemical

Skins can be softened from the underlying tissues by submerging vegetables in hot alkali solution. Lye may be used at a concentration of about 0.5-3%, at about 93° C (2000 F) for a short time period (0.5-3 min). The vegetables with loosened skins are then conveyed under high velocity jets of water which wash away the skins and residual lye. In order to avoid enzymatic browning, this chemical peeling is followed by a short boiling in water or an immersion in diluted citric acid solutions. It is more difficult to peel potatoes with this method because it is necessary to dissolve the cutin and this requires more concentrated lye solutions, up to 10%.

Thermal

Wet heat (steam): Other vegetables with thick skins such as beets, potatoes, carrots and sweet potatoes may be peeled with steam under pressure (about 10 at) as they pass through cylindrical vessels. This softens the skin and the underlying tissue. When the pressure is suddenly released, steam under the skin expands and causes the skin to puff and crack. The skins are then washed away with jets of water at high pressure (up to 12 at).

Dry heat (flame): Other vegetables such as onions and peppers are best skinned by exposing them to direct flame (about 1 min at 1000° C) or to hot gases in rotary tube flame peelers. Here too, heat causes steam to develop under skins and puff them so that they can be washed away with water.

Blanching Treatment

The special heat treatment to inactivate enzymes is known as blanching. Blanching is not indiscriminate heating. Too little is ineffective, and too much damages the vegetables by excessive cooking, especially where the fresh character of the vegetable is subsequently to be preserved by processing. This heat treatment is applied according to and depends upon the specificity of vegetables, the objectives that are followed and the subsequent processing / preservation methods.

Two of the more heat resistant enzymes important in vegetables are catalase and peroxidase. If these are destroyed then the other significant enzymes in vegetables also will have been inactivated. The heat treatment to destroy catalase and peroxidase in different vegetables are known, and sensitive chemical tests have been developed to detect the amounts of these enzymes that might survive a blanching treatment.

Because various types of vegetables differ in size, shape, heat conductivity, and the natural levels of their enzymes, blanching treatments have to be established on an experimental basis. As with sterilisation of foods in cans, the larger the food item the longer it takes for heat to reach the centre. Small vegetables may be adequately blanched in boiling water in a minute or two, large vegetables may require several minutes. Blanching as a unit operation is a short time heating in water at temperatures of 100° C or below. Water blanching may be performed in double bottom kettles, in special baths with conveyor belts or in modern continuous blanching equipment.

In order to reduce losses of hydrosoluble substances (mineral salts, vitamins, sugars, etc.) occurring during water blanching, several methods have been developed:

— temperature setting at 85-95° C instead of 100° C;
— blanching time has to be just sufficient to inactivate enzymes catalase and peroxidase;
— assure elimination of air from tissues.

Steam heat treatment can also be applied instead of water blanching as a preliminary step before freezing or drying, as long as the preservation method is only used for enzyme inactivation and not to modify consistency. For drying, the vegetables are conveyed directly from steaming equipment to drying installations without cooling. Vegetable steaming is carried out in continuous installations with conveyer belts made from metallic sieves.

Cooling of vegetables after water blanching or steaming is performed in order to avoid excessive softening of the tissues and has to follow immediately after these operations; one exception is the case of vegetables for drying which can be transferred directly to drying equipment without cooling. Natural cooling is not recommended because is too long and generates significant losses in vitamin C content. Cooling in pre-cooled air is sometimes used for vegetables that will be frozen

Cooling in water can be achieved by sprays or by immersion; in any case the vegetables have to reach a temperature value under 37° C as soon as possible. Too long a cooling time generates supplementary losses in valuable hydrosoluble substances; in order to avoid this, the temperature of the cooling water has to be as low as possible.

Canning Operation

Large quantities of vegetable products are canned. A typical flow sheet for a vegetable canning operation (which also applies to fruit for the most part) covers some food process unit operations performed in sequence: harvesting; receiving; washing; grading; heat blanching; peeling and coring; can filling;

removal of air under vacuum; sealing/closing, retorting/heat treatment; cooling; labelling and packing. The vegetable may be canned whole, diced, puréed, as juice and so on.

On-line Simplified Methods

Peroxidase Test

— *Solutions*: In order to check the peroxidase activity two solutions have to be prepared:

  — 1% guaiacol in alcohol solution (1 g guaiacol is dissolved in about 50 cm^3 of 96% ethylic alcohol and then this preparation is brought to 100 cm° with the same solvent);

  — peroxide solution 0.3% (1 cm^3 perhydrol is brought to 100 cm^3 with distilled water.

— *Sampling*: From various parts of the material samples are taken (about 20-30 pieces, etc.); the material is then crushed in a laboratory bowl in order to obtain an average sample.

— *Check*: Prom the average sample, 10-20 g of material is taken in a medium capacity test tube; on this sample are poured: 20 cm^3 distilled water; 1 cm^3 of 1% guaiacol solution; 1.6 cm^3 of peroxide solution.

The contents of the test tube is shaken well. The gradual appearance of a weak pink colour indicates an incomplete peroxidase inactivation—reaction slightly positive. If there are no tissue colour modifications after 5 minutes, the reaction is negative and the enzymes have been inactivated. As an orientative check it is also possible to simply pour a few drops of 1% guaiacol solution and 0.3% peroxide solution directly on blanched and crushed vegetables. A rapid and intensive brown-reddish tissue colouring indicates a high peroxidase activity (positive reaction).

Catalase Test

In order to identify the catalase enzyme activity, 2 g of dehydrated vegetables are well crushed and mixed with about 20 cm^3 of distilled water. After 15 min softening, 0.5 cm^3 of a 0.5% or 1% peroxide solution is poured on prepared vegetables. In the presence of catalase, a strong oxygen generation is observed for about 2-3 minutes. These tests are of a paramount importance in order to determine the vegetable blanching treatments (temperature and time); incomplete enzyme inactivation has a negative effect on finished product quality. For cabbage catalase inactivation by blanching is sufficient; blanching further to peroxidase inactivation would have negative effects on product quality and even complete browning. For all other vegetables and for potatoes, both tests must be negative, for catalase and for peroxidase.

Fresh Vegetable Storage

The vegetables can be stored, in some specific natural conditions, in fresh state, that is without significant modifications of their initial organoleptic properties. Fresh vegetable storage can be short term; this was briefly covered under temporary storage before processing. Also fresh vegetable storage can be long term during the cold season in some countries and in this case it is an important method for vegetable preservation in the natural state.

In order to assure preservation in long term storage, it is necessary to reduce respiration and transpiration intensity to a minimum possible; this can be achieved by:

— maintenance of as low a temperature as possible (down to 0° C),

— air relative humidity increased up to 85-95 % and

— CO_2 percentage in air related to the vegetable species.

Vegetables for storage must conform to following conditions: they must be of one of the autumn or winter type variety; be at edible maturity without going past this stage; be harvested during dry days; be protected from rain, sun heat or wind; be in a sound state and clean from soil; be undamaged. From the time of harvest and during all the period of their storage vegetables are subject to respiration and transpiration and this is on account of their reserve substances and water content.

The more the intensity of these two natural processes are reduced, the longer sound storage time will be and the more losses will be reduced. For this reason, vegetables have to be handled and transported as soon as possible in the storage conditions (optimal temperature and air relative humidity for the given species). Even in these optimal conditions storage will generate losses in weight which are variable and depend upon the species.

Vegetable Drying

Vegetable Powder Processing Technology

This technology has been developed in recent years with applications mainly for potatoes (flour, flakes, granulated), carrots (powder) and red tomatoes (powder). In order to obtain these finished products there are two processes:

— drying of vegetables down to a final water content below 4% followed by grinding, sieving and packing of products;

— vegetables are transformed by boiling and sieving into purées which are then dried on heated surfaces (under vacuum preferably) or by spraying in hot air.

Industrial installations that can be used for these products and technological data are summarised below:

— Dryers with plates under vacuum are equipped with plates heated with hot water. Stainless steel plates containing the purée to be dried are placed on them. Process conditions are at low residual pressure (about 10-20 mm Hg) and a product temperature of 50-70° C. This equipment is discontinuous but easy to operate.

— Drum dryers have one or two drums heated with hot water or steam as heating elements. Feeding is continuous between the two drums which are rotating in reverse direction (about 2-6 rotations per minute) and the distance of which is adjustable and determines the thickness of layer to be dried the product is dried and removed by mechanical means during rotation.

— Drying installations by spraying in hot air; the product is introduced in equipment and sprayed by a special device in hot air. Drying is instantaneous (1/50 s) and therefore can be carried out at 130-15O° C.

Packing and Storage of Dried Vegetables

Dried vegetables can suffer significant modifications that bring about their deterioration during storage. The factors that determine these degradations impose at same time the type of packaging materials and storage conditions for packaged products. The moisture content of dried vegetables is not constant because of their hygroscopicity and is always in equilibrium with relative humidity of air in storage rooms. Technical solutions for maintaining a low dehydrated products moisture are:

— storage in stores with air relative humidity below 78%;

— use packages that are water vapour proof. The most efficient packages are tin boxes or drums (mainly for long term storage periods); combined packages (boxes, bags, etc.) from complexes (carton with metallic sheets, plastic materials, etc.) mainly for small packages. One solution for some dried vegetables may be the use of waterproof plywood drums.

Modern solutions are oriented not only to the maintaining product moisture during storage but also reducing this parameter by the use of desiccants (substances which absorb moisture) introduced in packages, hermetically closed. A desiccant in current use is calcium oxide. Granulated calcium oxide is introduced in small bags from a material which is permeable to water vapour but which does not permit the desiccant to escape into products. With desiccants, product moisture can be reduced to even below 4%, and this inhibits or reduces the biochemical and microbiological processes during storage.

Another factor that can deteriorate dried/dehydrated vegetables is atmospheric oxygen through the oxidative phenomena that it produces. In order to eliminate the action of this agent some packing methods under vacuum or in inert gases are in use, applied mainly for packing dried carrots in order to avoid beta-carotene oxidation in beta-ionone. In order to avoid the action of oxygen it is also possible to add ascorbic acid as antioxidant.

Sun or artificial light action on dehydrated vegetables generally causes discoloration which can be avoided by using opaque packaging materials. Dehydrated vegetable compression (especially for roots) to form blocks with a weight of 50-600 g, is practiced sometimes; it has as advantages the reduction of evaporation surface and contact with atmospheric oxygen and volume reduction.

Dehydrated vegetables are compressed at about 300 at. Compressed blocks are packaged in heat sealed plastic materials. Storage temperature has an important role because this reduces or inhibits the speed of all physico-chemical, biochemical and microbiological processes, and thus prolongs storage period. The storage temperature should be below 25° C; lower temperatures help maintain taste, colour and water rehydration ratio and also, to some extent, vitamin C.

Potato Crisp Processing

The most important steps involved in potato crisps processing are:

— Selecting, procuring and receiving potatoes

— Storage of potato stock under optimum conditions

— Peeling and trimming the tubers

- Slicing
- Frying in oil
- Salting or applying flavoured powders
- Packaging

Table 1. Moisture and shipping factors for some dehydrated vegetables

Product	*Form/cut*	*Moisture %*	*Weight kg/m³*
Bean (green)	20 nun cut	5	1.6
Bean (lima)		5	3.3
Beet	6 mm strips	5	1.6-1.9
Cabbage	6-12 mm shreds	4	0.7-0.9
Carrots	5-8 mm strips	5	3-5
Celery	Cut	4	
Garlic	Cloves	4	
Okra	6 mm slices	8	
Onion	Slices	4	0.4- 0.6
Pea (fresh)	Whole	5	3.4
Pepper (hot)	Ground	5	
Pepper (sweet)	5 mm strips	7	
Potato (Irish)	5-8 mm strips	6	2.9-3.2
	Diced	5	3.3-3.6
Tomato	7-10 mm slices	35	

Selection and Storage

It is important to select potatoes of high specific gravity since this characteristic indicates superior yield and lower oil absorption. It is even more important to select potatoes with low reducing sugar contents or to store them at temperatures conducive to the minimising of these substances. Sprouting and fungal damage must also be minimised by the storage conditions.

Peeling Operation

The ideal peeling operation should only remove a very thin outer layer of the potato, leaving no eyes, blemishes, or other material for later removal by hand trimming. It should not significantly change the physical or chemical characteristics of the remaining tissue. Preferably peeling should use small amount of water and result in minimal effluent; compromises will have to be made in all of these aspects of peeling. First, the potatoes are thoroughly washed, not only for sanitary reasons, but also to prevent dirt of grit from abrading the equipment the tubers will dater contact.

Washing may take place in streams, as the potatoes are being conveyed by water streams, or in equipment provided with means for scrubbing the potato with brushes or rubber rolls. In barrel-type washers, potatoes are cleaned by being tumbled and rubbed against each other and against the sides of the barrel while they are immersed in, or sprayed with, water. After washing, the potatoes are allowed to drain, usually on mesh conveyors, and they travel over an inspection belt where foreign material and defective tubers are removed.

The more common peeling methods are abrasion, lye immersion, and steam. Abrasion peelers which may be either batch or continuous, use disks or rollers coated with grit to grind away the potato surface. An important design feature is to ensure that all surfaces of the tuber are equally exposed to the rasping action.

The peel fragments are flushed out of the unit by water sprays. Such systems work best with uniform, round, undamaged potatoes. Some of the advantages of abrasion peelers are their simplicity, compactness, low cost, and convenience. They are particularly suitable for peeling potatoes intended for chipping, since they do not chemically alter the surface layers. About 10% of the original tuber weight is lost through abrasion peeling prior to chipping.

Slicing

The peeled potatoes are cut into slices from 1/15 to 1/25 in. by rotary slicers. Centrifugal force presses the tuber against stationary gauging shoes and knives. Thickness is varied, not only to meet consumer preferences, but also to fit the condition of the tubers and the frying temperature and time. Slices produced at any one time must be very uniform in thickness, however, in order to obtain uniformly coloured chips. Slices with rough or torn surfaces lose excess solubles from ruptured cells and absorb larger amounts of fat.

It is necessary to remove the starch and other material released from the cut cells from the surface of slices so that the slices will separate readily and completely during frying. The slices are washed in stainless steel wire-mesh cylinders or drums rotating in a rectangular stainless steel tank. After washing and an additional rinse in similar equipment, the potatoes may or may not be dried.

Frying

The capacity of the fryer is generally the limiting factor in the process line. Most manufacturers currently use continuous fryers but some batch equipment is still employed.

Modern continuous fryers have the following essential elements: (1) a tank of hot oil in which the chips are cooked; (2) a means for heating and circulating the oil; (3) a filter for removing particles from oil; (4) a conveyor to carry chips out of the tank; (5) a reservoir in which oil is heated for adding to the circulating frying oil and (6) vapour-collecting hoods above the tank. Temperatures normally used are from 350 to 375° F at the receiving end and 320 to 345° F at the exit end.

The oil used for deep-fat frying of potato chips has two functions:

— it serves as a medium for transferring heat from a thermal source to the tuber slices;

— it becomes an ingredient of the finished product.

Use of highly refined oil is of great importance in flavour and stability of the crisps. Flavour, texture, and appearance are affected both by the amount of oil absorbed and its characteristics as it exists in the crisp. Oils change continuously during the frying process but the heat abuse resulting from the crisp cooking is relatively mild. Temperatures rarely rise above 385° F at any point. Better control over crisp colour could be obtained if the final stage of moisture removal could be achieved without the browning reaction that always accompanies it in the frying process.

Crisps may be sorted for size after frying, with the larger crisps being diverted to the bulk packs and larger pouches and the smaller pieces used for vending machine packs and other individual service containers. Potato crisp sizing is also accomplished by separating the peeled potatoes into large and small sizes, which are then sliced and fried separately. The crisps are salted immediately after they leave the fryer. It is important that the fat be liquid at this point to cause maximum adherence of the granules.

Powders containing barbecue spices, cheese, or other speciality materials may be added at this point. The salt may contain added enrichment materials or antioxidants. After salting, the crisps pass on to a conveyor belt where they are visually inspected and off-colour material is removed. The crisps are allowed to cool before packaging, better adherence of salt and flavour powders is obtained. Some consumers prefer the hard, curled-up crisp that is characteristic of the hand-kettle type of operation.

The special flavour of the hand-kettle crisp is said to be due, at least partly, to the starch retained on the cut surfaces of the potato slices as a result of the omission of a washing process after slicing. Starch-covered slices tend to stick together in the fryer so it is necessary to use devices to prevent clumping. The principal factors affecting potato crisp acceptability are piece size, colour, and of course, flavour. These factors are controllable primarily by selection of the raw material, adjustment of processing conditions, and packaging.

Storage Stability

If the frying oil is stabilised and has not deteriorated through use, and if the packaging is opaque and has a low moisture vapour transmittance rate (MVTR), a shelf life of 4-6 weeks should be achieved when crisps are stored at temperatures of about 70° F. Once potato crisps are in the bag, the three forms of quality loss which have the greatest effect on consumer acceptance are breakage, absorption of moisture with loss of crispiness, and fat oxidation leading to development of rancid odours.

The mechanical abuse causing breaking of the crisps can be partially prevented by using stiff packaging material, making the package "plump" with contained air, and avoiding crushing in the shipping case. Absorption of moisture is prevented largely by proper choice of packaging material. Cellophane coated with various moisture barriers has proved to be a satisfactory pouch films for the relatively short shelf-life expected.

Light (especially fluorescent light) accelerates oxidation, so that opaque packaging material must be used to obtain maximum shelf-life. Potato crisps are considered commercially unacceptable when they have a moisture content above 3%, which is in equilibrium with a relative humidity of about 32%. The containers should have a high degree of resistance to moisture-vapour transfer. If pouches are used, foil-containing films are preferable, since they not only resist moisture-vapour transfer but reflect light.

Vegetable Juices

Vegetable juices are natural products constituted from cellular juice and a part of crushed pulp, from the tissues of some vegetables. These juices contain all valuable substances from the vegetables: vitamins, sugars, acids, mineral salts and pectic substances. The most important of these products is tomato juice; in a lower proportion there are also other juices.

Tomato Juice

This product is characterised not only by its organoleptical properties (taste, colour, flavour) but also by its vitamin content close to those of fresh tomatoes. Modem technology is oriented to a maximum maintenance of organoleptic properties and of vitamin content.

At same time, it is important to assure juice uniformity by avoiding cellulosic particle sedimentation. Juice stability is assured by a flash pasteurization which assures the destruction of natural micro-flora, while keeping the initial properties.

The modern technological flow-sheet covers the following main operations:

— *Pre Washing* is carried out by immersion in water, cold or heated up to 50° C (possibly with detergents to eliminate traces of pesticides). This operation is facilitated by bubbling compressed air in the immersion vessel/equipment.

— *Washing* is performed with water sprays, which in modern installations have a pressure of 15 at or more.

— *Sorting/control* on rolling sorting tables enables the removal of non-standard tomatoes—with green parts, yellow coloured, etc.

— *Crushing* in special equipment.

— *Preheating* at 55-60° C facilitates the extraction, dissolves pectic substances and contributes to the maintaining of vitamins and natural pigments. In some modern installations, this step is carried out under vacuum at 630-680 mm Hg and in very short time.

— *Extraction* of juice and part of pulp (maximum 80%) is performed in special equipment / tomato extractors with the care to avoid excessive air incorporation. In some installations, as an additional special care, a part of pulp is removed with continuous centrifugal separators.

— *De-aeration* under high vacuum of the juice brings about its boiling at 35-40° C.

— *Homogenisation* is done for mincing of pulp particles and is mandatory in order to avoid future potential product "separation" in two layers. Flash Pasteurisation at 130-150° C, time = 8-12 see, is followed by cooling at 90° C, which is also the filling temperature in receptacles (cans or bottles).

— *Closing of Receptacles* is followed by their inversion for about 5 to 7 minutes.

— *Cooling* has to be carried out intensely.

Full cans do not need further pasteurisation because the bacteria that have potentially contaminated the tomato juice during filling are easily destroyed at 90° C due to natural juice acidity. For bottles, it may be possible to avoid further sterilisation if the following conditions can be respected: washing and sterilising of receptacles, cap sterilisation, filling and capping under aseptic conditions, in a space with UV lamps. In so far as this is quite difficult to achieve it may be necessary to submit bottles to a pasteurisation in water baths.

The main characteristics of high quality tomato juice are:

— natural red colour;

— taste and flavour of fresh tomatoes;

— uniformity (without pulp sedimentation);

— total soluble solids: 6% minimum;

— total soluble substances (by refractometer): 5% minimum;

— vitamin C: 15 mg/100ml minimum.

In traditional processes it is recommended to:

— thoroughly wash and rinse the empty receptacles (including jar caps / covers and bottle crown corks) and then "sterilise" by keeping in boiling water for 30 mini

— add salt and lemon juice to the prepared receptacles just before filling;

— pasteurise closed glass receptacles (bottles or jars) according to conditions recommended in technological flow-sheets and which is summarised as follows:

Receptacle size	Pre-heating	Time of pasteurization
0.33 1	60° C	40 minutes
0.501	60° C	45 minutes
0.66 1	60° C	55 minutes
0.751	60° C	60 minutes
1.0 litre	60° C	70 minutes

Carrot Juice

This product represents an important dietetic product due to its high soluble pectin content. Technological flow-sheet is oriented to the maintaining of as high as possible a pectin content and covers the following steps:

— Pre-Washing
— Cleaning
— Washing
— Blanching in steam for 20 minutes
— Grating
— Pressing
— Juice In the pressed juice will then be incorporated 25% of grated carrot (non pressed)
— Homogenisation in colloidal mills
— Acidification with 0.25% citric or tartric acid
— De-Aeration
— Filling in receptacles (bottles or tinplate cans)
— Airtight Sealing
— Pasteurisation at 100° C for 30 minutes.

The main characteristics of a good quality carrot juice:

— uniformity (no separation in layers occurs during storage);
— good orange colour;
— pleasant taste, close to fresh carrot taste;
— total soluble solids: 12 %;
— total sugar content: 8%;
— beta-carotene: 1.3 mg/100 ml;
— soluble pectin: 0.4 %.

Red Beet Juice

The product is obtained following this technological flow-sheet: washing, cleaning, steam treatment / steaming (30-35 min at 1050 C), pressing, strain through small hole sieve, filling in receptacles, tight sealing / closing, sterilisation (25 min at 1 15° C). In order to improve taste, the juice is acidified with 0.3% citric or tartric acid.

Sauerkraut Juice

Sauerkraut juice is produced in some countries for its dietetic value (lactic acid and vitamin C content) and its refreshing taste. The juice which is the result of the fermentation of lactic acid from cabbage, mainly from sliced sauerkraut, is used.

The juice must be the result of a normal lactic fermentation, i.e. without butyric fermentation or other deterioration. A good quality juice must have an acidity of 1.4% lactic acid and a content of maximum 2.5% salt; this is obtained by the mixing of various sauerkraut qualities.

The collected juice (from sauerkraut production) is heated slightly in order to eliminate CO_2 gas and to obtain protein coagulation. Filtration of juice is the next technological step, followed by filling in receptacles, closing of receptacles and pasteurisation at 75-80° C for 4-5 minutes.

Concentrated Tomato Products

Tomato Paste

The product with highest production volumes among concentrated products is tomato paste which is manufactured in a various range of concentrations, up to 44% refractometric extract. Tomato paste is the product obtained by removal of peel and seeds from tomatoes, followed by concentration of juice by evaporation under vacuum.

In some cases, in order to prolong production period, it may be advisable or possible to preserve crushed tomatoes with sulphur dioxide as described under semiprocessed fruit "pulps". Technological flow-sheets run according to equipment/ installation lay-outs, which are especially designed for this finished product. Manufacturing steps fall into three successive categories:

— obtaining juice from raw materials;

— juice concentration and

— tomato paste pasteurisation.

Obtaining juice from raw material. Preliminary operations (pre-washing, washing and sorting / control) are carried out in the same conditions as for manufacturing of "drinking" tomato juice described above. Next operation is removal of seeds from raw tomatoes: tomato crushing and seed separation with a centrifugal separator. Tomato pulp is pre-heated at 55-60° C and then passed to the equipment group for sieving: pulper, refiner and superrefiner with sieves of 1.5 mm, 0.8 mm and 0.4-0.5 mm respectively in order to give the smoothest possible consistency to the tomato paste.

Juice is concentrated by vacuum evaporation, a technological step which in modern installations runs continuously, tomato paste from the last evaporation step being at the specified concentration. In continuous installations with three evaporation steps (evaporating bodies), the juice is submitted in step / body I to pasteurisation at 85-900 C for 15 min and this will determine the microbiological stability of finished product. Vacuum degree corresponding to this temperature is 330 mm Hg.

In evaporating bodies II and III, temperatures are around 42-46° C and vacuum at 680700 mm Hg. Juice concentration occurs gradually and continuously in the three evaporating bodies. The advantages of continuous concentration are as follows:

- the taste, colour, flavour, "shine" and consistency of tomato paste are improved because:
    - the real concentration is performed in evaporating bodies II and III at low temperatures (42-46° C) and
    - the whole concentration process time from the input of juice in body I until the output of paste from body III is of about 1 hour (for paste with 30-35% refractometric extract).
- production capacity is raised by about 30% as compared to discontinuous installations with the same evaporation surface;
- the steam consumption is reduced by 60% because heating of bodies II and II is done with vapours resulting from juice evaporation in body I (double effect); water and electricity consumptions are also reduced by 30-40 %.

Tomato paste pasteurisation assures the microbiological stability of the product. For this purpose, the paste coming out from concentration equipment is passed continuously and in a "forced" mode through a tubular pasteuriser from which it emerge at a temperature of 90-92° C.

Usual commercial tomato paste types are at concentrations of 24%, 28% and 32% refractometric extract. Sometimes it is possible to obtain a tomato paste with a concentration of 44% refractometric extract; for this purpose it is necessary to eliminate a part of cellulose from tomatoes, an operation performed in a separating turbine.

Tomato paste storage and preservation is carried out after packing which is done usually in drums, metallic cans or glass jars; some modern equipment has been developed for packing in aluminium bags. As far as the concentration of tomato paste is concerned it is not possible to reduce water content down to 30% which corresponds to a water activity aw of 0.700.75 (minimum limit of mould growing), it is necessary to take special measures.

Salt is not a preservative in itself but contributes to the lowering of water activity. In drums, the preservation of tomato paste with minimum 30% refractometric extract is carried out in two ways:

— the hot paste (about 90° C) flows directly from pasteurisation equipment into drums that have been previously steamed;

— the paste is cooled down to 30° C through a heat exchanger and is introduced into drums that have been previously steamed.

For preservation purposes, it is possible to add 3-8% salt.

Preservation with 3% salt must be carried out respecting the following criteria:

— processing of a healthy raw material;

— thorough washing and control;

— pasteurisation of concentrated paste and use of well prepared drums. Paste in drums has to be stored in cold storage rooms during the hot season.

Preservation in big metal cans of 5 and 10 kg capacity of tomato paste with a minimum of 30% refractometric extract can be achieved without sterilisation if the following conditions are respected:

— sterilisation by steam of cans and covers;

— filling of paste at 92-94° C;

— airtight sealing/ closing of cans;

— invert cans and then

— air cooling.

For small packages (tinplate cans of 1/10-1/1 or glass jars of same capacity) it is usual to use pasteurised paste, as hot as possible (92-94° C). The receptacles are first sterilised by steam. After airtight sealing, the receptacles are kept in boiling water for a short time in order to sterilise their inner surface and the paste in contact with inner receptacle surface. In some countries small receptacles are not further sterilised if the manufacturing is carried out in perfect hygienic and sanitary conditions.

Packing in small tinned aluminium tubes is carried out with concentrated paste, pasteurised and hot. Good quality tomato paste is an homogenous mass, with a high density, without foreign bodies (seeds, peel, etc.), with a red colour, and an agreeable taste and smell, close to those of fresh tomatoes. There are usually three types of tomato paste: 36, 30 and 24 which have refractometric extracts of respectively 34-38%, 28-32% and 24-26%. Paste of good quality must have a volatile acidity of maximum 0.15% as lactic acid. An 8% salt addition is accepted.

Concentrated Tomato Juice

Concentrated tomato juice is a product with 17-19 % refractometric extract and is a homogenous mass, finely sieved, without foreign bodies / and without any evidence of deterioration. A good quality product has a red colour, an agreeable and specific taste and smell. Modern technology uses the same installations, equipment and flow-sheets for concentrated tomato juice as for the production of tomato paste; the final concentration is thus regulated between the above specified limits.

The concentrated tomato juice is filled in receptacles (metal tinplate cans or glass bottles) and then pasteurised at 100° C during 15-25 minutes according to receptacle type. With modern production lines it should be possible to pass the concentrated tomato juice through a tubular pasteuriser and then pack aseptically and cool, without the need to pasteurise the receptacles.

Tomato Sauces

Under the USA Code of Federal Regulation 7 CFR 52, 1991 tomato sauce is the concentrated product prepared from the liquid extract from mature, sound, whole tomatoes, the sound residue from preparing such tomatoes for canning, or the residue from partial extraction of juice, or any combination of these ingredients, to which is added salt and spices and to which may be added one or more nutritive sweetening ingredients, a vinegar or vinegars, and onion, garlic, or other vegetable flavouring ingredients. The refractive index of the tomato sauce at 20° C is not $<$ 1.3461.

These products are widespread in some countries and are used in order to spice some meals. Sauces can be obtained from fresh tomatoes or from concentrated products, those from fresh tomatoes being of superior quality. Technological processing covers the following steps: concentrated juice processing, addition of flavour/taste ingredients, boiling, fine sieving, filling of receptacles, closing and pasteurisation. Tomato sauces which can be sweet, more or less spicy are prepared according to specific recipes.

Production Accidents and Product Defects

Tomato Juice

- "Separation" in layers is due to not enough homogenisation or low / insufficient viscosity. In the first case it is necessary to intensify homogenisation; and in second to increase the pre-heating temperature to 60° C in order to obtain protopectine hydrolisis and pectolitic enzymes inactivation.
- Moulding of the juice is brought about by significant infection of raw materials, inadequate washing and control or by use of contaminated packages. The preventive measures should be decided after cause analysis.
- Fermentation of juice is manifested by a significant development of gases. Prevention methods are the same as for moulding.
- Tomato juice turns sour, without the formation of gases; this defect is initiated by thermophyl and thermoresistant bacteria; the juice acquires a vinegary taste. Prevention: maintenance of flash pasteurisation temperature at 130-135° C.
- Excessive vitamin C losses are due to a simultaneous action of heating and oxygen from air. Prevention:
  - prevent air going into crusher and extractor;
  - assure an intensive de-aeration (vacuum degree 700 mm Hg) at a temperature of at least 35-40° C; and
  - close receptacles in vacuum.
  - Weak colour of tomato juice can be avoided by the utilisation of mature tomatoes and with a pulp of as red a colour as possible.

Tomato Paste and Concentrated Juice

- Presence of sand is caused by inadequate washing or by a significant contamination of raw material; this can be prevented by a more intensive pre-washing and washing of tomatoes.

— There may be mould especially at the surface of tomato paste packed in drums. Prevention:
  — accurate pre-washing and washing;
  — follow pasteurisation instructions;
  — pack in clean drums or receptacles; and
  — close receptacles immediately after filling.
— Fermentation is manifested by a weak alcohol smell or by a weak vinegar taste; when the fermentation is more advanced there is gas production in the product mass. Prevention: as for moulding prevention.

Tomato Sauces

Surface of the product turns black at the contact zone with air; this is due to the action of iron on the tannins from spices, tomato seeds, etc. Prevention:

— avoid iron equipment;
— avoid crushing of tomato seeds and
— seal receptacles in vacuum.

Pickles and Sauerkraut Technology

Natural Acidification Technology

Gherkins and Cucumbers

Raw materials must follow strict specifications for a high quality finished product; the following parameters must be considered as critical:

— adapt a uniform size according to the finished product requirements; for example, gherkins will need to have a maximum length of 9 cm for raw vegetables. Generally 15 cm size/length will be a maximum for high quality cucumber products in many countries. However, according to local preferences, bigger cucumbers could be also in demand.
— cylindrical or ovoidal shape;
— dark green colour;
— absence of surface defects due to cryptogamic diseases.

Cucumbers have to be picked at their ripeness for eating, when the sugar content is at about 1.5-2.2%, needed for lactic fermentation. Unripe cucumber does not have enough sugar.

The general technological flow-sheet is as follows:

— Reception

— Control

— Temporary Storage

— Grading By Size

— Washing

— Small Holes are made in large size cucumbers skin;

— Receptacle Filling: raw material is simply put in the receptacles in bulk, with care to arrange them in such a way that a maximum of pieces could be introduced;

— Salt Solution Preparation: 6% salt solution (NaCl);

— Salt Solution Addition: the salt solution is poured into the receptacle;

— Fermentation is carried out at 20-30° C, anaerobically. This step takes generally 4-8 weeks. Acidity reaches a value up to 1.5% lactic acid (and in some exceptional cases up to 2% lactic acid) which corresponds to a maximum pH value of 4.1.

— Storage; after the last fermentation stage, drums and other receptacles have to be stored at low temperature; best conditions for 12 months shelf life should be below + 15° C. Storage temperature will determine the shelf life of the products.

— Addition of 1000 ppm potassium sorbate will prevent mould development without having any influence on lactic fermentation.

— Raw material grading by size is a very important technological step. In order to accelerate brine penetration, mainly for medium to large size cucumbers, the practice of making small holes in the raw material skins is generally recommended.

— A major factor influencing the quality of lactic fermented cucumbers is the water durity; optimal results are obtained at 15-20° durity.

Cucumber consistency / texture is influenced by the formation of calcium pectate with the pectic substances from raw material tissues. In some countries, calcium chloride (0.3-0.5 %) is added in order to firm up the cucumber consistency. Chlorinated water which still contains active chlorine can inhibit or even stop the lactic fermentation.

Sauerkraut

In some countries cabbages are submitted to lactic fermentation as whole vegetables; however, in many countries the cabbage is shredded before fermentation. As shredded cabbage and its technology is at the basis of an important industry, giving good quality products, with a uniform fermented product and with good keeping quality and ease of distribution, this will be described first.

Cabbage as raw material for sauerkraut must be sound, ripe for eating, well-leafed and from suitable varieties. Optimum total sugar level needed for the lactic fermentation is 24%; generally good quality raw material contains up to 30-60 mg/100 g of vitamin C.

Shredded Sauerkraut

The technological flow sheet is as follows:

— Reception

— Control

— Temporary Storage is carried out in bulk, up to a height of about 1 m, during few days. This step produces a heat generation which facilitates later fermentation by the softening of tissues.

— Removal of External Leaves

Coring is done with a specially adapted mechanical screw; this operation generates small particles of finely divided cabbage which will be mixed with the main part of vegetable during shredding / chopping. The core represents about 10% from the whole cabbage, is rich in sugar and vitamin C, but being too high in fibre content needs to be chopped separately as described. Shredding/Cutting of cabbage is carried out with complex specific equipment which is generally installed directly on the "top" of fermentation silos and is mobile, installed on rails and moves all along the silos. The dimension of resulting shredded cabbage is about 2-3 mm thick.

The same complex equipment is designed to grind the added salt to fine particles and to distribute shredded cabbage and ground salt in an uniform manner to the fermentation silos. The usual capacity of fermentation silos is up to 30 tons, with separate compartments of 45 tons each. Salt Addition is carried out by the equipment described above; the proportion of salt is 2-2.5% with respect to cabbage. This proportion must not be changed because the salt in this technology does not have a preservative role but only that to extract from cabbage the juice needed for fermentation. It is be preferable to obtain a fairly light pressure on cabbage just after salt addition with some simple mechanical means. This is important in order to:

— create an anaerobic medium for fermentation;

— facilitate external diffusion of cellular juice;

— assure a rational use of the fermentation space.

Fermentation: The maximum acidity level obtained is generally of about 1.5% lactic acid (and very rarely 2.5%); this is obtained in 4-6 weeks. Optimal acidity is 1.0-1.8% and pH value 4.1 or lower. Fermentation temperature is at 20-25° C in the first phase and needs to be lowered then to 14-18° C. During fermentation, the brine from each storage / fermentation silo cell is periodically circulated with a pump in order to uniformise the fermentation process.

Storage is performed in same silos used for fermentation, or the finished products is removed from silos and packed in drums and other receptacles according to distribution schedule. These silos are usually made of reinforced concrete and coated with gritstone plates or with an acid-resisting material layer. At small scale and in traditional processing, shredded sauerkraut can be obtained by using simple available glass or rigid plastic receptacles. At home, this process can use glass jars and / or local / traditional pottery receptacles from a minimum size of 2-3 kg up to the available / practical sizes. In some countries shredded sauerkraut is preserved in receptacles by pasteurization, once the fermentation process has been completed.

Whole Sauerkraut

According to the consumer preference in different countries and to the specific situations it is also usual to preserve whole cabbages by lactic fermentation. At small or medium scale operations, whole cabbage could be processed/ fermented in cylindric receptacles like 30 to 200 litre rigid plastic drums, or rectangular receptacles made from food grade rigid plastic. It is possible to find this type of drum in a significant number of developing countries. These two types of rigid plastic receptacles could also be used for shredded sauerkraut production.

Prepared whole cabbages are put into fermentation receptacles and a 5-6 % salt concentration brine is poured on top. The fermentation conditions are the same as for shredded sauerkraut. In order to assure a uniform fermentation and to avoid a strict anaerobic (butyric) fermentation it is necessary to apply a periodic juice "aeration".

Other Acidified Vegetables

In principle all vegetables with a sugar content of at least 2 % could be preserved by lactic fermentation. From a practical point of view it is mainly the following vegetables which are preserved by this technology: unripe tomatoes (green tomatoes), peppers, eggplant, carrots and cauliflower, alone or usually in a mix with cucumber as mixed pickles.

Fermentation of individual vegetables is carried out according to a flow-sheet as described for whole sauerkraut. The type of cut, brine concentration and frequency of operating steps have to be adapted to each case; green tomatoes are fermented as whole vegetable.

Artificial Vegetable Acidification Technology

This technology is based on the addition of food grade vinegar which has a bacteriostatic action in concentrations up to 4 % acetic acid and bactericidal action in higher concentrations. Vegetables preserved in vinegar need to reach, after equilibrium between vinegar and water contained in vegetables, a final concentration of 2-3 % acetic acid in order to assure their preservation. To achieve this final concentration, a 6-9 % acetic acid vinegar is used, as related to the specific ratios vinegar/ vegetables. In vinegar pickles, salt (2-3 %) and sometimes sugar (2-5 %) are also added. If the vinegar concentration is lower than 2%, vinegar pickles need to be submitted to a pasteurisation in order to assure their preservation.

Cucumbers in Vinegar

This represents the basic product obtained by this technology. Cucumbers have to be wholesome, with a soft texture and not have reached eating maturity. They must have a low sugar content because in this technology there is no lactic fermentation involved. Dimensions are up to 12 cm length, with a preference for small cucumbers.

The technological steps are the followings:

- Size Grading
- Washing
- Arrange In Receptacles—glass jars, etc.

— Pouring Of Vinegar is usually carried out at room temperature; however, hot vinegar addition enables a sterilisation of cucumber surface and facilitates vinegar penetration in vegetable tissues.

— Salt (Sugar) Addition

— Spicing Addition

— The technological cycle of artificial acidification is considered completed when acetic acid concentration reaches an equilibrium value; the time needed is about 2 weeks.

When equilibrium concentration in acetic acid is below 2 %, the cucumbers are submitted to a pasteurisation for 20 min at 90-1000 C in order to assure their preservation. Cucumbers in vinegar with previous lactic fermentation are excellent quality products because the lactic fermentation improves the taste of these cucumbers. The principle of this process is to assure preservation both by acetic acid and by lactic acid simultaneously.

Technological processing flow-sheet is as follows: small cucumbers ("cornichons" or "gherkins") are washed, brushed and small holes are made in the skin; the vegetables then are put in drums with slightly warm 6% brine which also contains spices. The lactic fermentation runs for few days up to a lactic acid concentration of about 0.5 %. The cucumbers are removed from the brine, washed thoroughly and well drained.

Preservation is usually done in glass jars by pouring a normally flavoured vinegar with about 9% acetic acid usually in order to bring the final concentration to 3% calculated as acetic acid. In order to obtain a high quality product only wine vinegar should be used. In some pickles (e.g. in "Cornichons") the usual level of wine vinegar is set at 20 % of packaged product total weight; some alcohol vinegar could be still added and final concentration will be adjusted as described above.

Other Vinegar Pickles

One type in this category is represented by other vegetables acidified with vinegar separately or in a mix (red peppers, sweet green pepper, green tomatoes, cauliflower, etc.). The preparation steps are similar to the ones used for cucumbers in vinegar. Significant quantities of special mixed vegetables in vinegar are manufactured in many countries, with the international name of "mixed pickles" with following composition: small cucumbers ("cornichons"/"gherkins")—maximum 70 mm in length -, sliced carrots, cauliflower, small onions (less than 25 mm diameter), mushrooms etc. and spices.

The vegetables are acidified separately in vinegar and then are put into receptacles (glass jars); a flavoured vinegar, salted and sweetened with acetic acid concentration of 3-5% is poured over them. In the case of lower acetic acid concentrations, a pasteurisation at 90° C for 10-20 minutes is applied according to the receptacle size.

Canned Vegetable

Canned vegetables can be classified as follows:

- canned products in salt brine;
- canned products in tomato concentrated juice;
- canned products in vegetable oil.

Canned Vegetables in Tomato Juice

General technological flow-sheet covers two types of operations:

- Preparation of vegetables is similar to the one described for canned vegetables in salt brine: sorting, washing, grading, cutting, blanching and cooling; the exception is for spices which are not blanched.
- Preparation of canned products covering: receptacles filling with vegetables, adding concentrated tomato juice (with minimum 8% refractometric extract), hermetic closing/sealing of receptacles, sterilisation and cooling of receptacles.

One usual composition for mixed vegetables in tomato juice is:

- eggplants (slices): 20%
- peppers (cut): 20%
- carrots (slices): 15%
- green peas: 5%
- green beans (pods): 18%
- okra (whole): 8%
- tomatoes (whole or halves): 14%

Each vegetable is prepared separately as in general canning operation description. At receptacle filling for mixed vegetables products, each vegetable should be introduced separately in specified proportions; hot concentrated tomato juice (at least 700 C) is poured onto the vegetables.

Heat Preservation Operations Canning

The success of heat preservation operations lies in:

— selecting suitable fruit and vegetables in good conditions;

— preparing them hygienically and skilfully;

— packing them in cans which are hermetically sealed and then processed under fixed conditions of time and temperature;

— cooling these cans carefully and storing them under conditions which will not cause deterioration of either the cans or their contents.

Selection of Raw Materials

It is appreciated that some varieties of fruit and vegetables are not suitable for canning, either because they are uneconomical to prepare or because the colour, flavour or texture are poor. Suitable varieties must be available to the canner in quantities sufficient to meet his requirements and in sound conditions for canning.

The flow to the cannery should be regulated in order that perishable materials are not left for a long time before being handled, since any delay will cause deterioration. Apart from the main ingredients, be it fruit or vegetables, minor ingredients also require careful selection. Sugar, salt, water and spices for instance may all be contaminated with spoilage organisms, so constant testing of all raw materials is essential.

Preparation

This is carried out by various methods, including grading, trimming, peeling, washing and blanching. All equipment must be scrupulously clean and preparation should be completed quickly and carefully in order to keep the bacterial load as low as possible. Thorough washing of vegetable is necessary to remove spores of heat resistant bacteria which are present in large numbers in the soil. Blanching in steam or hot water is of no avail against these heat resistant spores because of the comparatively low temperatures involved.

Reasons for blanching are:

— the removal of gas from the tissues of the raw material;

— the shrinkage of this material;

— the inhibition of enzymic reactions, which, if not checked, will adversely affect the colour and nutritive value of the food.

Filling

Filling, be it mechanical or by hand, requires careful attention. The cans must be clean and the correct weight of foodstuffs must be added. Under-filled cans will be underweight and the headspace will be too large, resulting in too much air being left in the can. Overfilling may lead to seams being strained during processing and to ends becoming distorted and bulged. If the product forms hydrogen on storage as is the case with coloured fruits, swelling of the can due to hydrogen pressure will occur more quickly in an overfilled can than in one which has been correctly filled. Overfilling also affects heat penetration in the can and may lead to spoilage outbreaks.

Air Removal

Before the can is seamed, air must be removed from the contents and the headspace. Normally, this is carried out by passing the cans through a steam box until the temperature at the centre of the can is at least 160° F. This operation, termed exhausting, is necessary for the following reasons:

— to minimise strains on the seams due to expansion of air during the processing period;

— to remove oxygen which accelerates corrosion in the can and also causes oxidation of the food with possible serious effects on colour and flavour;

— to reduce the destruction of vitamin C;

— to enable a vacuum to be formed when the can is cooled.

This ensures that the ends remain concave, even when storage temperatures are a little higher than usual, and also acts as a reservoir for hydrogen which may be formed by reactions between the can and its contents. Thus a high vacuum makes for a long shelf life. Large cans, however, should not reach such a high exhaust temperature before seaming as smaller cans because of the danger of the can body collapsing on cooling, a condition known as "panelling".

Double Seaming

The can should be double-seamed as soon as the correct centre temperature has been attained. Any delay between exhausting and seaming will lead to loss of vacuum and may lead to bacterial spoilage. The quality of the double seam must, of curse, be frequently checked.

Heat Processing

After seaming, the cans are heated for a definite time at a definite temperature to kill or inhibit organisms which may cause spoilage. This operation is termed "heat processing". The times and temperatures required for "heat processing" of various packs have been determined experimentally to ensure that spores of the most heat resistant food poisoning organisms known, Clostridium botulinum, are destroyed. There are other organisms, however, whose spores are more heat resistant than those of Clostridium botulinum and which although they will not cause food poisoning may cause spoilage and for this reason the minimum heat processing time is often exceeded by recommendations made by laboratories.

At the same time there is a limit to the amount of heating which a canned food may be given without spoiling its flavour, texture and colour and this also has to be taken into consideration when process recommendations are made. Bacterial spores have a greater resistance to heat when the growth-medium is neutral or near neutral, and neutrality is normally required for bacterial growth to commence. Because of this, canned foods have been broadly divided into two groups:

— "acid" foods having a pH of 4.5 or lower and

— "non-acid" foods having a pH of more than 4.5.

"Non-acid" foods (vegetables) must, therefore be "heat processed" at high temperatures using steam under pressure, whereas "acid" foods (fruit) may be processed at the (lower) temperature of boiling water, since this will kill moulds and yeasts and if any bacterial spores survive the combination of acid and heat, they will be inhibited from growth by the acid environment. The rate of destruction by heat follows a definite pattern, the same proportion of the surviving bacteria being destroyed in successive units of time.

The more bacteria there are in a pack, the more time will be need to reduce their numbers. For this reason, it is essential that the initial number of bacteria be kept low, and this may be achieved by ensuring fast and hygienic handling at all stages in the cannery. Pressure gauges and retort temperature control equipment must be checked frequently for accuracy. Processing times and temperatures must be strictly adhered to, and complete removal of air from the retort during processing must be achieved by adequate venting. Failure to remove the air completely will result in their being cold spots in the retort and intermittent spoilage is likely.

Cooling

As soon as the heat processing time is completed, the cans are cooled in chlorinated water as rapidly as possible without damaging them. Cans processed in steam develop high internal pressure because of the expansion of the foodstuff, the expansion of air in the can and the increase in the vapour pressure of the water in the can.

During the heat process, these pressures are counter-balanced to some extent by the pressure of the steam in the retort, but on releasing this steam pressure at the commencement of the cooling period, the pressure in the can may be sufficient to strain the seams seriously and may even distort the ends. Cans of A21/2 size or larger, when processed at temperatures of 240° F or more, are liable to undergo permanent distortion, such as peaking.

This may be avoided by pressure cooling, which involves replacing steam pressure by air pressure before introducing water to the retort, and maintaining this until the pressure inside the can has fallen to a safe level. This presents difficulties, since if the air pressure is maintained after the can has developed a vacuum, the can body is liable to collapse. Where pressure-cooling is not carried out, the retort pressure is allowed to drop slowly to atmospheric pressure and the cans are then cooled with water.

Storage

After cooling, the cans should be stored in cool, dry conditions. The maintenance of a constant temperature is desirable, since a rise in temperature may lead to condensation of moisture on the can, with possible rusting. Cool conditions are required because storage at higher temperatures not only causes chemical and physical changes in the product and the container but also introduces a risk of thermophilic spoilage.

Other known causes of container spoilage in storage are the use of labels and cardboard cases which have too high a chloride content, and the use of unseasoned wood in the manufacture of packing cases, all of which tend to cause rust formation on the cans.

Quality Control

The international trade in processed fruits and vegetables is very large with an ever increasing number of different types being processed and exported. Whereas once, processing was limited to mostly temperate climate fruits and vegetables, the change has now broadened to include tropical and subtropical types.The reasons are twofold. Firstly, consumers' dietary habits have become more diverse so that, for example people living in North America may very well like fruit and vegetables grown in Africa or Asia. Secondly, processing techniques, whether they be for canning, freezing or drying, have been improved to an extent where final product is palatable, nutritious and of long and reliable shelf life.

Many developing countries have taken advantage of the continuing worldwide demand for processed fruits and vegetables and earned valuable foreign exchange from exports of products to profitable markets.

The export quality control and inspection of processed fruits and vegetables is directed at ensuring that the final products:

— have been processed in a registered export establishment that is constructed, equipped and operated in an hygienic and efficient manner;
— conform to the requirements of the export regulations for processed fruits and vegetables, and those of the importing country, in respect of such things as quality grades, defects, ingredients, packaging materials, styles, additives, contaminants, fill of container, drained weight; and,
— conform to labelling requirements

References

Biaugeaud, H. 1994. Food processing equipment. *Technical Bulletin*. Henri Biaugeaud, S.A.: Arcueil.

Colin, D. 1992. Recent trends in fruit and vegetable processing. In *Food Science and Technology Today*, 7, (2)), pp. 111 - 116

FAO. 1990. *Rural processing and preserving techniques for fruits and vegetables*. Rome: FAO.

Toledo, R.T., 1991. *Fundamentals of Food Process Engineering*, 2nd Edition, van Nostrand, Reinhold, New York.

4

Fruit Processing

Fruit is sometimes defined as the product of growth from an angiosperm, or flowering plant. From a purely botanical point of view, the fruit may be only the fleshy growth that arises from the ovary of a flower and may not necessarily include any other structures. From the consumer's or food processor's point of view, however, fruit is generally characterized as the edible product of a plant or tree that includes the seed and its envelope and can typically be described as juicy, sweet, and pulpy.

Fruits are a high-moisture, generally acidic food that is relatively easy to process and that offers a variety of flavour, aroma, colour, and texture to the diet. They are usually low in calories but are an excellent source of dietary fibre and essential vitamins. Owing to the presence of cellulose, pectin, and various organic acids, fruits can also act as natural laxatives. Fruits are therefore a valuable part of the diet.

Nutrient Composition

Fresh fruit is typically between 75 and 95 percent water, a fact that helps to explain the refreshing character of the food. In general, fruits are acidic, with pH ranging from 2.5 to 4.5. The most common acids in fruits are citric acid, malic acid, and tartaric acid.

Of all the vitamins present in fruits, the most noted is vitamin C, or ascorbic acid. Actual quantities of vitamin C in fruits are not especially large, but the vitamin is particularly important in the diet because of its role in the prevention of disease and in the general promotion of good health. Citrus fruits, such as oranges, lemons, and grapefruits, are well known for their vitamin C content. Other sources include most berries and melons. Carotene, a chemical common to fruit, is easily converted in the body to vitamin A; cantaloupes, peaches, and apricots are significant sources of this nutrient.

Typically, fruits are high in carbohydrates, although a large range is possible—between 2 and 40 percent, depending on the type of fruit and its maturity. Free sugars usually include fructose, glucose, and sucrose; other sugars may be present in smaller quantities.

A large portion of the carbohydrates present in fruits is fibre, which is not digested and passes through the digestive system. Fibre is usually made up of cellulose, hemicellulose, and pectic substances. A small amount of starch may also be present in fruit, but starches are typically converted to sugars during the ripening process.

A negligible quantity of protein is found in fruits, and they usually contain less than 1 percent fat. Fats are most typically associated with the waxy cuticle surface of the fruit skin. Exceptions to this rule are avocados and olives, the flesh of which may contain as much as 20 percent oil.

Maturation and Spoilage

Fruits are living biological entities that perform a number of metabolic functions. Two functions of particular importance in fruit processing are respiration (the breaking down of carbohydrates, giving off carbon dioxide and heat) and transpiration (the giving off of moisture). Once the fruit is harvested, respiration and transpiration continue, but only for as long as the fruit can draw on its own food reserves and moisture. It is this limited ability to continue vital metabolic functions that defines fruit as perishable.

Fruit development can generally be divided into three major stages: growth, maturation, and senescence. The period of growth generally involves cell division and enlargement, which accounts for the increasing size of the fruit. Maturation is usually reached just prior to the end of growth and may include flavour development and increase in sugar content. Senescence is the period when chemical synthesizing pathways give way to degradative processes, leading to aging and death of tissue. Fruit ripening is thus the result of many complex changes, some interactive but many independent of one another.

As harvested fruit ages, it is particularly important to manage the temperatures under which it is stored. For example, respiration largely involves enzymatic processes, which are significantly controlled by ambient temperature. The rate of chemical change in fruit generally doubles for every increase of 10°C (at room temperature, roughly 20° F).

Changes that take place during storage as fruit begins to overripen may include extreme colour formation, development of strong off-flavours with intense aroma, softening of the flesh, onset of physiological disorders, and manifestations of disease. In addition, fruit can be injured by overcooling. Chilling injury may be manifested by pitting and browning of the surface and by pitting and darkening of the flesh.

Microorganisms can also cause problems during senescence and storage. Many bacteria and fungi, for instance, are involved in decay after harvest. Typical fungi include Alternaria, Botrytis, Monilinia, Penicillium, and Rhizopus. These fungi are generally weak pathogens, in that they usually invest only weak or damaged fruit. Efforts to control infection begin in the orchard, usually with the application of fungicides. Cooling of the fruit or, conversely, hot-water dipping may also enhance storage quality. In addition, the careful application of ionizing radiation has been shown to inhibit microbial growth.

Once harvested, fruits are moved to storage. In the case of highly heat-sensitive products such as raspberries or cherries, the fruit should be precooled prior to storage. Precooling can be accomplished by hydrocooling (immersion of the fruit in cold water) or vacuum cooling (moistening and then placing under vacuum in order to induce evaporative cooling).

A typical storage system for fruit is cold storage, using refrigerated air. Other techniques include controlled-atmosphere (CA) storage and hypobaric storage. In CA storage the oxygen and carbon dioxide content of the storage environment are controlled in such a way as to retard senescence and further deterioration of the fruit. In general, oxygen levels are reduced and carbon dioxide levels increased. CA conditions can be generated in a number of ways. Conventional CA depends on the respiration of the fruit to generate carbon dioxide, and the concentration of this gas is controlled by wet scrubbers, hydrated lime, or other commercial carbon dioxide removal systems. Liquid nitrogen and compressed nitrogen gas have also been used to flush out the ambient air of the storage facility. In other systems oxygen is converted to carbon dioxide by reaction with liquid propane or by catalytic burning.

Hypobaric storage involves the cold storage of fruit under partial vacuum. Typical conditions include pressures as low as 80 and 40 millimetres of mercury and temperatures of 5° C (40° F). Hypobaric conditions reduce ethylene production and respiration rates; the result is an extraordinarily high-quality fruit even after months of storage.

Harvesting and Preprocessing

There is a distinction between maturity and ripeness of a fruit. Maturity is the condition when the fruit is ready to eat or if picked will become ready to eat after further ripening. Ripeness is that optimum condition when colour, flavour and texture have developed to their peak. Some fruit is picked when it are mature but not yet ripe. This is especially true of very soft fruit like cherries and peaches, which when fully ripe are so soft as to be damaged by the act of picking itself.

Further, since many types of fruit continue to ripen off the tree, unless they were to be processed quickly, some would become overripe before they could be utilised if picked at peak ripeness. From a technological point of view, fruit characterisation by species and varieties is performed on the basis of physical as well chemical properties: shape, size, texture, flavour, colour/ pigmentation, dry matter content (soluble solids content), pectic substances, acidity, vitamins, etc. These properties are directly correlated with fruit utilisation.

The proper time to pick fruit depends upon several factors; these include variety, location, weather, ease of removal from the tree (which change with time), and purpose to which the fruit will be put. For example, oranges change with respect to both sugar and acid as they ripen on the tree; sugar increases and acid decreases. The ratio of sugar to acid determines the taste and acceptability of the fruit and the juice. For this reasons, in some countries there are laws that prohibit picking until a certain sugar-acid ratio has been reached. In the case of much fruit to be canned, on the other hand, fruit is picked before it is fully ripe for eating since canning will further soften the fruit.

Many quality measurements can be made before a fruit crop is picked in order to determine if proper maturity or degree of ripeness has developed. Colour may be measured with instruments or by comparing the colour of fruit on the tree with standard picture charts. Texture may be measured by compression by hand or by simple type of plungers. As fruit mature on the tree its concentration of juice solids, which are mostly sugars, changes. The concentration of soluble solids in the juice can be estimated with a refractometer or a hydrometer.

The refractometer measures the ability of a solution to bend or refract a light beam which is proportional to the solution's concentration. A hydrometer is a weighted spindle with a graduated neck which floats in the juice at a height related to the juice density. The acid content of fruit changes with maturity and affects flavour. Acid concentration can be measured by a simple chemical titration on the fruit juice. But for many fruits the tartness and flavour are really affected by the ratio of sugar to acid.

Percentage of soluble solids, which are largely sugars, is generally expressed in degrees Brix, which relates specific gravity of a solution to an equivalent concentration of pure sucrose. In describing the taste of tartness of several fruits and fruit juices, the term "sugar to acid ratio" or "Brix to acid ratio" are commonly used. The higher the Brix the greater the sugar concentration in the juice; the higher the "Brix to acid ratio" the sweeter and lees tart is the juice.

The above and other measurements, plus experience, indicate when fruit is ready for harvesting and subsequent processing. A large amount of the harvesting of most fruit crops is still done by hand; this labour may represent about half of the cost of growing the fruit. Therefore, mechanical harvesting is currently one of the most active fields of research for the agricultural engineer, but also requires geneticists to breed fruit of nearly equal size, that matures uniformly and that is resistant to mechanical damage.

A correct manual harvesting includes some simple but essential rules:

— the fruit should be picked by hand and placed carefully in the harvesting basket; all future handling has to be performed carefully in order to avoid any mechanical damage;

— the harvesting basket and the hands of the harvester should be clean;

— the fruit should be picked when it is ready to be able to be processed into a quality product depending on the treatment which it will undergo.

It is worth emphasising the fact that the proximity of the processing centre to the source of supply for fresh raw materials presents major advantages; some are as follows:

— possibility to pick at the best suitable moment;

— reduction of losses by handling/transportation;
— minimises raw material transport costs;
— possibility to use simpler/cheaper receptacles for raw material transport.

Once it has left the tree, the organoleptic properties, nutritional value, safety and aesthetic appeal of the fruit deteriorates in varying degrees. The major causes of deterioration include the following:

— growth and activity of micro-organisms;
— activities of the natural food enzymes;
— insects, parasites and rodents;
— temperature, both heat and cold;
— moisture and dryness;
— air and in particular oxygen;
— light and
— time.

The rapidity with which foods spoil if proper measures are not taken is indicated in Table 1.

Table 1. Useful storage life of some food products

Food Products (Days) at 21°C (70°F)	Generalised Storage Life
Animal Flesh, Fish, Poultry	1-2
Dried, salted, smoked meat and fish	360 and more
Fruits	1-7
Dried Fruits	360 and more
Leafy Vegetables	1-2
Root Crops	7-20

Fruit Reception

Fruit reception at the processing centre is performed mainly for following purposes:

— checking of sanitary and freshness status;
— control of varieties and fruit wholeness;
— evaluation maturity degree;

— collection of data about quantities received in connection to the source of supply: outside growers/farmers, own farm.

Variety control is needed in order to identify that the fruit belongs to an accepted variety as not all are suitable for different technological processes. Fruit maturity degree is significant as industrial maturity is required for some processing/preservation methods while for others there is the need for an edible maturity when the fruit has full taste and flavour. Special attention is given to size, appearance and uniformity of fruit to be processed, mainly in the form of fruit preserved with sugar using whole/half fruits ("with fruit pieces").

Some laboratory control is also needed, even if it not easy to precisely establish the technological qualities of fruit because of the absence of enough reliable rapid analytical methods able to show eventual deterioration. The only reliable method for evaluating the quality is the combination of data obtained through organoleptic/taste controls and by simple analytical checks which are possible to perform in a small laboratory: percentage of soluble solids by refractometer, consistency/texture measured with simple penetrometers, etc.

Temporary Storage Before Processing

This step has to be as short as possible in order to avoid flavour losses, texture modification, weight losses and other deterioration that can take place over this period.

Some basic rules for this step are as follows:

— keep products in the shade, without any possible direct contact with sunlight;

— avoid dust as much as possible;

— avoid excessive heat;

— avoid any possible contamination;

— store in a place protected from possible attack by rodents, insects, etc.

Cold storage is always highly preferred to ambient temperature. For this reason a very good manufacturing practice is to use a cool room for each processing centre; this is very useful for small and medium processing units as well. The type of analysis for audits will be adapted to the specific fruits and vegetables that are received/ processed. An excellent indication of a good temporary storage is the limited weight loss before processing, which has to be below 1.0%-1.2%.

Washing Efficiency

Harvested fruit is washed to remove soil, micro-organisms and pesticide residues. Fruit washing is a mandatory processing step; it would be wise to eliminate spoiled fruit before washing in order to avoid the pollution of washing tools and/or equipment and the contamination of fruit during washing. Washing efficiency can be gauged by the total number of microorganisms present on fruit surface before and after washing—best result are when there is a six fold reduction. The water from the final wash should be free from moulds and yeast; a small quantity of bacteria is acceptable. Fruit washing can be carried out by immersion, by spray/ showers or by combination of these two processes which is generally the best solution: pre-washing and washing.

Some usual practices in fruit washing are:

— addition of detergents or 1.5% HCl solution in washing water to remove traces of insect-fungicides;

— use of warm water (about 50°C) in the pre-washing phase;

— higher water pressure in spray/shower washers.

Washing must be done before the fruit is cut in order to avoid losing high nutritive value soluble substances (vitamins, minerals, sugars, etc.).

Fruit Sorting

Fruit sorting covers two main separate processing operations:

— removal of damaged fruit and any foreign bodies;

— qualitative sorting based on organoleptic criteria and maturity stage.

Mechanical sorting for size is usually not done at the preliminary stage. The most important initial sorting is for variety and maturity. However, for some fruit and in special processing technologies it is advisable to proceed to a manual dimensional sorting (grading).

Trimming and Peeling of Fruit

This processing step aims at removing the parts of the fruit which are either not edible or difficult to digest especially the skin. Up to now the industrial peeling of fruit and vegetables was performed by three procedures:

— mechanically;
— by using water steam;
— chemically; this method consists in treating fruit and vegetables by dipping them in a caustic soda solution at a temperature of 90 to 100° C; the concentration of this solution as well as the dipping or immersion time varying according to each specific case.

Cutting

This step is performed according to the specific requirements of the fruit processing technology.

Heat Blanching

Fruit is not usually heat blanched because of the damage from the heat and the associated sogginess and juice loss after thawing. Instead, chemicals are commonly used without heat to inactivate the oxidative enzymes or to act as antioxidants and they are combined with other treatments.

Ascorbic Acid Dip

Ascorbic acid or vitamin C minimises fruit oxidation primarily by acting as an antioxidant and itself becoming oxidised in preference to catechol-tannin compounds. Ascorbic acid is frequently used by being dissolved in water, sugar syrup or in citric acid solutions. It has been found that increased acidity also helps retard oxidative colour changes and so ascorbic acid plus citric acid may be used together. Citric acid further reacts with (chelates) metal ions thus removing these catalysts of oxidation from the system.

Sulphur Dioxide Treatment

Sulphur dioxide may function in several ways:

— sulphur dioxide is an enzyme poison against common oxidising enzymes;
— it also has antioxidant properties; i.e., it is an oxygen acceptor (as is ascorbic acid);
— further SO_2 minimises non enzymatic Maillard type browning by reacting with aldehyde groups of sugars so that they are no longer free to combine with amino acids;
— sulphur dioxide also interferes with microbial growth.

In many fruit processing pre-treatments two factors must be considered:

— sulphur dioxide must be given time to penetrate the fruit tissues;

— SO_2 must not be used in excess because it has a characteristic unpleasant taste and odour, and international food laws limit the SO_2 content of fruit products, especially of those which are consumer oriented.

Commonly a 0.25 % solution (except for semi-processed fruit products which are industry oriented and use a 6% solution) of SO_2 or its SO_2 equivalent in the form of solutions of sodium sulphite, sodium bisulphite or sodium/potassium metabisulphite are used. Fruit slices are dipped in the solution for about two to three minutes and then removed so as not to absorb too much SO_2. Then the slices are allowed to stand for about one to two hours so that the SO_2 may penetrate throughout the tissues before processing.

Sulphur dioxide is also used in fruit juice production to minimise oxidative changes where relatively low heat treatment is employed so as not to damage delicate juice flavour. Dry sulphuring is the technological step where fruit is exposed to fumes of SO_2 from burning sulphur or from compressed gas cylinders; this treatment could be used in the preparation of fruits (and some vegetables) prior to drying / dehydration.

Sugar Syrup

Sugar syrup addition is one of the oldest methods of minimising oxidation. It was used long before the causative reactions were understood and remains today a common practice for this purpose. Sugar syrup minimises oxidation by coating the fruit and thereby preventing contact with atmospheric oxygen. Sugar syrup also offers some protection against loss of volatile fruit esters and it contributes sweet taste to otherwise tart fruits. It is common today to dissolve ascorbic acid and citric acid in the sugar syrup for added effect or to include sugar syrup after an SO_2 treatment.

Fresh Fruit Storage

Some fruit species and specially apples and pears can be stored in fresh state during cold season in some countries' climatic conditions. Fruit for fresh storage have to be autumn or winter varieties and be harvested before they are fully mature. This fruit also has to be sound and without any bruising; control and sorting by quality are mandatory operations.

Sorting has to be carried out according to size and weight and also by appearance; fruit which is not up to standard for storage will be used for semi-processed product manufacturing which will be submitted further to industrial processing. Harvested fruit has to be transported as soon as possible to storage areas. Leaving fruit in bulk in order to generate transpiration is a bad practice as this reduces storage time and accelerates maturation processes during storage.

Fruit Drying and Dehydration Technology

For fruit with a high sugar content drying temperatures have to be lower at initial stage and then increase to the maximum acceptable; for fruit with lower sugar level the temperatures are applied in a reverse order.

Processing of Fruit Bars

The fruit bar processing method developed for FAO only involves a single major operation, which it drying the fruit pulp after mixing it with suitable ingredients. It can be used to produce mango, banana, guava or mixed fruit bars. A dual-powered dryer, working by solar energy during the day and by electric or steam power at night and on rainy days, with cross-flow movement of air and controlled temperature (from 55° C at the beginning of processing to a high of 70° C), is well suited for dehydration of the pulp to the desired moisture level of 15 to 20%.

Mango fruit bar: Fully ripe mangoes are selected and washed in water at room temperature. The peeled fruit is cut into slices and passed through a helicoidal pulper to extract the pulp. The required amount of sugar to adjust the Brix (the unit measure for total solids in fruits) of the mixed pulp to 25 degrees Brix is then added. Two grams of citric acid per kilogram of pulp (or 20 ml of lime or lemon juice) are added to inhibit possible growth of microorganisms during drying. The mixture is then heated for two minutes at 80° C and partially cooled; the heat treatment serves to inactivate the enzymes and destroy the microorganisms.

Potassium or sodium metabisulphite is added (two grams per kg of prepared mixture), so that the concentration of SO_2 is 1000 ppm. The mixture is then transferred to stainless steel trays which have been previously smeared with glycerine (40 ml/m^2). Each tray must be loaded with 12.5 kg of mixture per square metre.

Drying could be carried out by a dual-powered dryer for a total of 26 hours:

— 10 hours by solar energy at about 55° C and

— 16 hours by electric or steam power at 70° C.

At the end of the drying operation, when moisture content is between 15 and 20%, the pieces of suitable shape and size are wrapped in cellophane paper, packed in cartons and stored at ambient air temperature. Pieces of unsuitable shape and size are further cut into small pieces and used to prepare, along with peanuts and cashews, a variety of "cocktail mixtures".

Banana fruit bar: Banana varieties which give smooth pulp without serum separation must be used for this purpose. Ripe, suitable fruit is selected. The hand-peeled fruits are soaked in 0.3 per cent citric acid solution for about 10 minutes (lime or lemon juice can replace citric acid). The drained fruit are pulped to obtain smooth pulp.

Guava fruit bar: A mixture of pink and yellow varieties is best suited for preparing the bar. The washed fruit is hand peeled and stem and blossom ends trimmed. The peeled fruit is cut into quarters which are passed through a helicoidal extractor to separate seeds and fibrous pieces (the holes in the stainless steel screen should be between 0.8 and 1.10 mm). To get the maximum yield of pulp, the material is passed through the extractor twice. After adjusting the refractometric solids to 25 degrees Brix, the fruit bar can be prepared by following the same procedure as for mango pulp.

Mixed fruit bar: Mango and banana pulp, as well as papaya and banana pulp, can be mixed in a calculated ratio for preparing mixed fruit bars. The rest of the procedure is the same as in the case of pure mango pulp.

Packing and storage: The dried pulp is removed from the dryer and cut into square pieces of 5 x 5 cm at a thickness of about 0.3 cm. These pieces, arranged in three layers make up blocks of about 0.9 cm thickness weighing between 25 and 28 grams. An unit pack consist of two such blocks and weights between 50 and 56 grams. Each block is separately wrapped in cellophane and the unit pack is filled in a printed cellophane bag of size 15 x 8 cm. Two hundred unit packs are packed in a master carton of size 34 x 22 x 14 cm, with a net weight of about 10 kg. Shelf-life is about one year at room temperature.

Fruit leathers: Fruit leathers are manufactured by drying/dehydration of fruit purées into leathery sheets. The leathers are eaten as confections or cooked as a sauce. They are made from a wide variety of fruits, the more common being apple, apricot, banana, cherry, blackcurrant, grape, peach, pear, pineapple, plum, raspberry, strawberry, kiwi fruit, mango and papaya.

A product with good potential is ciku leather; ciku fruit is grown in Malaysia. A standard process is carried out using ripe fruits which are washed, peeled, diced and the seeds removed. The fruits are blanched for 1 minute at 80° C and blended into puree in a food processor. Ciku leather is prepared by mixing ciku puree with 10% sugar, 10% pre-gelatinous rice flour, 150 ppm sorbic acid an 500 ppm sodium metabisulphite ($Na_2H_2SO_4$).

The mixture is cooked on a water bath at 60° C and then made into sheets 1.8 mm thick on trays spread with glycerol to reduce stickiness. This is then further dried in a forced-air dehydrator at 45° C for 3.5 hr or until the surface no longer feels sticky when touched with the fingers. The dried and cooled leathers are cut into 12 x 12 cm squares and wrapped in polypropylene (PP) of 0.1 mm thickness.

Osmotic Dehydration

Osmotic dehydration is a useful technique for the concentration of fruit and vegetables, realised by placing the solid food, whole or in pieces, in sugars or salts aqueous solutions of high osmotic pressure. It gives rise to at least two major simultaneous counter-current flows: a significant water flow out of the food into the solution and a transfer of solute from the solution into the food.

Process Variables

Main process variables are:

— pre-treatments;

— temperature;

— nature and concentration of the dehydration solutions;

— agitation;

— additives.

In the light of the published literature, some general rules can be noted:

— water loss and solid gain are mainly controlled by the raw material characteristics and are certainly influenced by the possible pre-treatments;

— it is usually not worthwhile to use osmotic dehydration for more than a 50% weight reduction because of the decrease in the osmosis rate over time. Water loss mainly occurs during the first 2 hr and the maximum solid gain within 30 min.;
— the rate of mass exchanges increases with temperature but above 45 ° C enzymatic browning and flavour deterioration begin to take place. High temperatures, i.e. over 60° C, modify the tissue characteristics so favouring impregnation phenomena and thus the solid gain;
— the best processing temperature depends on the food; mass exchanges are favoured by using high concentration solutions;
— phenomena which modify the tissue permeability, such as over-ripeness, pre-treatments with chemicals (SO_2), blanching or freezing, favour the solid gain compared to water loss because impregnation phenomena are enhanced;
— the kind of sugars utilised as osmotic substances strongly affects the kinetics of water removal, the solid gain and the equilibrium water content. Low molar mass saccharides (glucose, fructose, sorbitol, etc.) favour the sugar uptake;
— addition of NaCl to osmotic solutions increases the driving force for drying.

Synergistic effects between sugar and salt have also been observed.

Applications

The effects of osmotic dehydration as a pre-treatment are mainly related to the improvement of some nutritional, organoleptic and functional properties of the product. As osmotic dehydration is effective at ambient temperature, heat damage to colour and flavour is minimised and the high concentration of the sugar surrounding fruit and vegetable pieces prevents discoloration.

Furthermore, through the selective enrichment in soluble solids high quality fruit and vegetables are obtained with functional properties "compatible" with different food systems. These effects are obtained with a reduced energy input over traditional drying process. The main energy-consuming step is the reconstitution of the diluted osmotic solution that could be obtained by concentration or by addition of sugar.

Drying

Air drying following osmotic dipping is commonly used in tropical countries for the production of so-called "semi-candied" dried fruits. The sugar uptake, owing to the protective action of the saccharides, limits or avoids the use of SO_2 and increases the stability of pigments during processing and subsequent storage period. The organoleptic qualities of the end product could also be improved because some of the acids are removed from the fruit during the osmotic bath, so a blander and sweeter product than ordinary dried fruits is obtained. Owing to weight and volume reduction, loading of the dryer can be increased 2-3 times.

The combination of osmosis with solar drying has been put forward, mainly for tropical fruit. A 24 hour cycle has been suggested combining osmodehydration, performed during the night, with solar drying during the day. Two-three-fold increase in the throughput of typical solar dryers is feasible, while enhancing the nutritional and organoleptic quality of the fruits. A two-step drying process, OSMOVAC, for producing low moisture fruit products was described.

The osmotic step is performed with sucrose syrup 65-75 Brix until the weight reduction reaches 30-50%. By osmotic dehydration followed by vacuum drying puffy products with a crisp, honeycomb-like texture can be obtained at a cost comparatively lower than freeze-drying. Commercial feasibility of the process on bananas has been studied, based on the results of a semi-pilot scale operation. Osmotically dried bananas retained more puffiness and a crisper texture than simple vacuum dried ones, and the flavour lasted longer at ambient temperature. The combination of osmotic dehydration with freeze-drying has been proposed only at laboratory scale.

Appertisation

A combination of osmotic dehydration with appertisation has been proposed to improve canned fruit preserves. The feasibility of a process, called osmo-appertisation, to obtain high quality fruit in syrup, has been assessed on a pilot scale. The key point of this technique is the pre-concentration of the fruit to about 20-40 Brix, that causes, together with the enhancement of the natural flavour, an increase of the resistance of the fruit to the following heat treatment, especially for colour and texture stability.

The products obtained are stable up to 12 months at ambient temperature and show a higher organoleptic quality than canned preserved alternatives. Furthermore, because of their higher specific weight and diminished volume, the filling capacity of jars or pouches is increased.

Freezing

The frozen fruit and vegetable industry uses much energy in order to freeze the large quantity of water present in fresh products. A reduction in moisture content of the material reduces refrigeration load during freezing. Other advantages of partially concentrating fruits and vegetables prior to freezing include savings in packaging and distribution costs and achieving higher product quality because of the marked reduction of structural collapse and dripping during thawing.

The products obtained are termed "dehydro-frozen" and the concentration step is generally carried out through conventional air drying, the additional cost of which has to be taken into account. Osmotic dehydration could be used instead of air drying to obtain an energy saving or a quality improvement especially for fruit and vegetable sensitive to air drying.

Juice Extraction

An osmotic pre-step before juice extraction was reported to give highly aromatic fruit or vegetable juice concentrates.

Further Developments

So far only applications on a pilot plant scale are reported in the literature. For further developments on a larger scale, theoretical and practical problems should be solved. The industrial application of the process faces engineering problems related to the movement of great volumes of concentrated sugar solutions and to equipment for continuous operations. The use of highly concentrated sugar solutions creates two major problems. The syrup's viscosity is so great that agitation is necessary in order to decrease the resistance to the mass transfer on the solution side. The difference in density between the solution and fruit and vegetables, makes the product float. Another important aspect, so far not investigated, is the microbiological safety of the process, which should be studied thoroughly before further industrial development.

Osmoappertisation

In order to obtain an alternative to the canned fruit preserves and to maintain a high quality of the fruits, a research has been carried out on the osmoappertisation of apricots, a "combined" technique that consists in the appertisation of the osmodehydrated apricots. This technique could contributes also to the reduction of energy consumption, limits the cost of production and combines "convenience" (ready-to-eat, medium shelf-life) with many market outlets (retail, catering, bakery, confectionery, semi-finished products). Osmoappertisation combines two unit operations: dehydration by osmosis and appertisation (packaging + pasteurization).

Apricot Processing

Fresh Apricot Purée

After washing, cutting and removal of stones, apricot halves are dipped in 2% solution of sodium or potassium metabisulphite for 10 minutes. After draining, the resulting material is passed through a 0.045-in. screen pulper—finisher to produce a fresh apricot puree. The fresh apricot puree obtained in this way could be further processed in different semiprocessed (i.e. chemically or otherwise preserved products) or finished fruit products (fruit leathers, fruit bars, jams, etc.).

Concentrated Apricot Pulp

Fresh apricot halves could also be steam blanched for 5 min., passed through a 0.045-in pulper—finisher and transformed in a purée with about 14 Brix depending on the fruit quality. This purée may be concentrated in steam jacketed kettles up to 20 Brix or in other adequate equipment (e.g. a stirred vacuum evaporator) up to 28° Brix. As for fresh apricot purée, the concentrate may be further processed in various semiprocessed or finished fruit products as mentioned above and as will be described below.

Dried Apricot Leather

- From fresh fruit purée by drum drying. The fresh apricot purée at about 14 Brix could be dried to 12% moisture apricot sheets, using a double-drum dryer operating at 132 degrees C with a drum clearance of 0.008 in and speed of 45 sec per revolution.
- From fruit concentrate by drum drying. The concentrate could also be dried to 12% moisture fruit sheets by the same process as described above.
- From fresh fruit purée or from apricot concentrate by sun/solar drying or by dehydration.

Trays: For sun / solar drying or dehydration of fruit pulp, the trays must have a solid base in order to retain the liquid contents. They may be made of metal, timber or plastic. Stainless steel or plastic trays are most suitable because they are unaffected by acid fruit pulp; they are, however, expensive. A metal tray could be 75 x 50 cm in size and with side 5 cm high. The trays must keep level during drying; if the tray is not level the pulp will run to the lowest point, giving a layer of irregular depth which will dry unevenly.

Any tray which is not made of stainless steel or plastic must be covered inside with a sheet of heavy gauge plastic film to protect the pulp from chemical or bacteriological contamination. Standard sun/solar trays as described can be used by covering them inside with a sheet of plastic film to create a solid base.

Preparation before drying/dehydration: Fresh apricot purée can be directly used for next processing steps. Fruit concentrate needs to be added to potassium metabisulphite to obtain a 0.3% concentration of SO_2 in the material.

Drying/dehydration: The apricot/fruit purée or concentrate is poured into the trays to a depth of about 1.5 cm. When stainless steel or plastic trays are used they should be coated with a thin layer of glycerine to prevent sticking. The pulp is then sun/solar dried or tunnel/cabinet dehydrated; moisture content in the dried product should not exceed 14% and the SO_2 content should not be less than 1500 pp.

The dried product is wrapped in cellophane to prevent sticking, then put inside polythene bags and stored at best in tight fitting tins and sealed to prevent moisture transfer. From fresh fruit purée or from apricot concentrate, with sugar addition, and then processed by sun / solar drying or by dehydration. In some countries preference is for finished products with added sugar; and this is also interesting from a point of view of energy consumption (concentration is partially achieved by sugar dry matter) and of shelf life. The overall content in SO_2 could also be reduced as sugar is a preservation agent, the product will be close to a fruit "paste".

Reconstitution Test

In reconstitution water is added to the product which is restored to a condition similar to that when it was fresh. This enables the food product to be cooked as if the person was using fresh fruit or vegetable. All vegetables are cooked but many of the dried fruits can be used for eating after they have been soaked in water. The following reconstitution test is used to find out the quality of the dried product.

— Weigh out a sample of 35 grams from the bulked and packed final product of the previous day's production.

— Put the sample into a small container (beaker) and add 275 ml of cold water (and 3.5 g salt).

— Cover the container (with a watch-glass) and bring the water to the boil.

— Boil Gently for 30 minutes.

— Turn out the sample onto a white dish.

— At least two people should then examine the sample for palatability, toughness, flavour and presence or absence of bad flavours. The testers should record their results independently.

— The liquid left in the container should be examined for traces of sand/ soil and other foreign matter.

This test can be used also to examine dried products after they have been stored for some time. Evaluation of rehydration ratio may be performed according to the following calculations.

Rehydration ratio: If weight of the dried sample is 10 g (Wd) and the weight of the sample after rehydration is 60 g (Wr), rehydration ratio is:

$$\frac{Wr}{Wd} = \frac{60}{10} = \frac{6}{1}, \text{or 6 to 1}$$

Rehydration coefficient: The weight of rehydrated sample is 60 g (Wr); the weight of dried sample is 10 g (Wd) and its moisture is 5% (Wu); raw material before drying had 87% water (A); rehydration coefficient is:

$$\frac{Wr}{Wd - Wu} = \frac{60}{\quad} = \frac{780}{9.5} = 82.1$$

A simpler test for eating quality can be carried out without weighing and measuring. The material is placed in a cooking pot with water (and a little salt). The pot is then covered and boiled as described above. Except for a few products which are eaten in the dry state, most dried fruit and all dried vegetables are prepared by soaking and cooking. Often this preparation is carried out incorrectly and dried products get a bad reputation.

Good quality dried products, after cooking and if properly treated should be similar to cooked fresh produce. In order to get good results, the following methods are recommended:

— *Quick method*: Cold water, ten times the weight of the dry product, is added to the dried product. The container is covered, brought to the boil and simmered GENTLY until the product is tender. The cooking time may be 15 to 45 minutes after the boiling point has been reached.

— *Slow method*: This gives better results than the quick method. Cold water is added to the dry food and is left to soak for 1 to 2 hours before cooking. The product is then cooked in the same water as that in which it was soaked. The actual cooking time will probably be shorter than that for the quick method.

Other points to remember are:

— if too much water is added the cooked product will have little flavour. However, if too little water is added the product may dry and burn. This can be avoided by adding small quantities of water during cooking;

— always cook with a lid on the container;

— salt, if required, should be added when the cooking is almost complete;

— partly used packages of dry products should be reclosed tightly or kept in containers with good fitting lids.

Handling, Sorting, Packing and Storage of Fruits

It is not easy to assess when drying has been completed. In absence of instrumentation, the characteristics of the various products after drying / dehydration can only be assessed by experience. Although this cannot be conveyed adequately on paper, some general indications can be given.

Fruit products: When a handful of fruit is squeezed tightly together in the hand and then released, the individual pieces should drop apart readily and no moisture be left behind on the hand. It should not be possible to separate the skin by rubbing unpeeled fruit and the fruit centre should no longer reveal any moist area. Banana should be leathery and not too tough to eat in their dry state.

Vegetable products: Onions should be dried until they are crisp whereas tomatoes should be leathery. In general, the lower the moisture content, the better the keeping quality will be, but overdried products generally have an inferior quality. Also the loss in weight from excessive drying cannot be tolerated in a commercial operation designed to run profitably. It is, however, essential to dry up to an optimum / safe moisture level, related to the type of the product and his designed shelf life, and to avoid running the risk of the products becoming spoiled due to excess water content.

When drying is completed, the material should be sorted either on trays or on a table in order to remove pieces of poor quality and colour and any foreign matter. Very fine material should be separated from the bulk of the material by using a sieve (12 or 16 mesh per inch). Bad quality products which show poor colour need to be removed from the bulk of finished product. After selection and grading, dried products should be packed immediately, preferably in polythene bags which must be folded and closed / tied tightly. However, plastic bags are easily damaged and therefore they must be packed into cartons or jute sacks before they are transported.

Deterioration of Dried Fruit

Dried fruit must be considered as a relatively perishable commodity in the same category as cereals, pulses and similar stored products. It is subject to deterioration resulting from mould growth, insect and mite infestation and physical and chemical changes.

Mould Growth

When the moisture content of dried fruit is allowed to exceed the maximum permissible level for safe storage then mould growth may occur. At the present time, suitable field moisture meters for use with dried fruit are not readily available, and moisture determinations can only be satisfactorily carried out where laboratory facilities are available. Various species of drought resisting fungi may develop on dried fruit when the moisture content is just above the safe level, and a number of osmophilic yeasts are quite commonly associated with spoilage in dried fruit.

Many of the yeasts bring about fermentation with the production of lactic acid or alcohol, and yeasts are frequently present in wart-like crystalline growths which occur in fruit which has become "sugared". In very moist fruit mucoraceous fungi may predominate and are visible as white fluffy growths on and within the fruit.

Mite Infestation

Severe mite infestations are often associated with the growth of osmophilic yeasts in fermenting dried fruit products. Many of these mites are unable to complete their development in the absence of yeast. They have been reported as occurring on dried fruit, and particularly figs and prunes in Mediterranean countries. Such infestations are difficult to eradicate and affect consumer acceptance of the contaminated products.

Insect Infestation

Insect infestations may begin in the field before harvest, may continue during bulk storage after drying, and unless measures are taken to prevent it, may occur in the finished packaged product during storage prior to distribution and consumption. Regular treatments of the stack of dried fruit with a suitable insecticide will be necessary as a routine to combat light insect infestations. Pyrethrins synergised with piperonyl butoxide are commonly used as a surface spray or as an aerosol fog for this purpose. Heavy infestations will require that the fruit be fumigated.

Technology of Semi-processed Fruit Products

The semi-processed fruit products are manufactured in order to be delivered to industry processing centres (in the fruit producing country itself or in importing countries) where they will be further manufactured in consumer oriented finished products: jams, jellies, syrups, fruits in syrup, etc.

In the practice of semi-processed fruit products and for the purpose of this chapter the following categories are defined:

- *fruit "pulps"*: semiprocessed products, not refined, obtained by mechanical treatment (or, less often, by thermal treatment) of fruit followed by their preservation. Either whole fruit, halves or big pieces are used which enables easy identification of the species. "Pulps" can be classified in boiled or non boiled (raw).
- *fruit "purées-marks"*: semiprocessed products obtained by thermal and mechanical treatment or, very rare, raw and then refined, operations by which all nonedible parts (cores, peels, etc.) are removed. "Purées-marks" are classified in boiled (the more usual case) and non boiled (raw).
- *semiprocessed juices*: products obtained by cold pressure or very rare by other treatments (diffusion, extraction, etc.) followed by the preservation.

Preservation of Semiprocessed Fruit Products

Preservation can be achieved by chemical means, by freezing or by pasteurization. The choice of preservation process for each individual case is a function of the semiprocessed product type and the shelf life needed.

Chemical Preservation

In many countries, in practice, this is carried out with sulphur dioxide, sodium benzoate, formic acid and, on a small scale, with sorbic acid and sorbates. Preservation with sulphur dioxide is a widespread process because of its advantages: universal antiseptic action and very economic application. The drawbacks of SO_2 are: SO_2 turn firms the texture of some fruit species (pomaces), desulphiting is not always complete and recolouring of red fruits is not always complete after desulphitation. Practical preservation dosage levels with SO_2 for about 12 months is 0.18-0.20% SO2 (with respect to the product to be preserved). This level could be reduced to 0.09% SO_2 for 3 months and to 0.12% SO_2 for 6 months preservation. The preservation with sulphur dioxide is in use mainly for "pulps" and for "purées-marks".

Chemical preservation can be performed from a practical point of view by the utilisation of 6% SO_2 water solutions or by direct introduction of sulphur dioxide gas in the product (for "purées-marks"). The preparation of 6% SO_2 solutions is done by bubbling the gas from cylinders in cold water; from a 50 kg SO_2 compressed gas cylinder results 830 l of 6% SO_2 solution. These SO_2 solutions have to be stored in cool places, in closed receptacles and with periodic concentration control/check by titration or by density measurements approximate results.

Preservation with sodium benzoate has the following advantages: it does not firm up the texture and does not modify fruit colour. The disadvantages are: it is not a universal antiseptic, its action needs an acid medium and the removal is partial. Sodium benzoate is in use for "pulps" and for "purées-marks" but less for fruit juices. Practical dosage level for 12 months preservation is 0.18-0.20 % sodium benzoate, depending on the product to be preserved. Sodium benzoate is used as a solution in warm water; the dissolution water level has to be at maximum 10% reported to semiprocessed product weight.

Formic acid preservation is performed mainly for semiprocessed fruit juices at a dosage level of 0.2 % pure formic acid (100%). Formic acid is an antiseptic effective against yeasts, does not influence colour of products and is easily removed by boiling. Formic acid could be diluted with water in order to insure a homogeneous distribution in the product to be preserved;

water has to be at maximum 5 % of the product weight. Because of a potential effect of pectic substance degradation, formic acid is less in use for "pulps" and "purées-marks" preservation.

Sorbic acid used as potassium sorbate can be used for preservation of fruit semiprocessed products at a dosage level of 0. 1% maximum. Advantages of sorbates are: they are completely harmless and without any influence on the organoleptic properties of semiprocessed fruit products.

Preservation by Pasteurisation

As fruit has a low pH, preservation of semiprocessed fruit products could also be performed by pasteurisation, the length of this step varying with the size of the receptacles. The advantages of this type of treatment are: hygienic process, which assure a long term preservation; the disadvantages are: need for air tight receptacles, and pectic substances could begin to deteriorate if the thermal treatment is too long.

Thermal preservation of fruit semiprocessed products could also be done by a "selfpasteurization": very hot semiprocessed products are filled into receptacles which are sealed and then inverted in order to sterilise the air which goes through the hot fruit mass.

Preservation by Freezing

This is done on an industrial scale in some countries and can be done with or without sugar addition. The advantages of this process are: absence of added substances; very good preservation of quality of fruit constituents (pectic substances, vitamins, etc.) and good preservation of organoleptic properties (flavour, taste, colour). Freezing is done at about -20 to -30° C and storage at -10 to -18° C. Freezing is applied mainly to semiprocessed fruit products aimed at very high quality and high cost finished products.

Chemical Preservation

Sorting is needed in order to remove sub-standard fruit (with moulds, with diseases, etc.) and all foreign bodies.

Washing is obligatory in order to remove all impurities which cannot be eliminated at the processing step in finished products.

Coring and Cutting, mainly for pomace fruits, has as main objective a better utilisation of preservation "space" in receptacles and is not mandatory; this will be defined by customer/ supplier agreements / standards. This operation is preferably performed by mechanical means.

Preservation is carried out with the 6% SO_2 solution which is added to the prepared fruits (placed in bulk in receptacles) in the quantity needed to obtain the preservation dosage level. For a better / homogeneous preservative distribution, the initial 6% SO_2 solution could be diluted with water; however, the diluted solution (which will be filled in receptacles) has to be at a dosage level of less than 10% of the semiprocessed product weight.

For some soft fruit, especially strawberries, preservation is done with a mix of 6% SO_2 solution and calcium bisulphite solution (containing also 6% SO_2). Preparation of calcium bisulphite solution is done by the introduction of 30 kg of CaO in 1 m^3 SO_2 solution and mixing up to clarification. The resulting solution is mixed with the initial 6% SO_2 solution, generally in a 1:1 ratio, but the ratio can be adapted to the fresh fruit texture. Firming of soft fruit texture by this treatment is based on the formation of calcium pectate with pectic substances from fruit tissues.

In the case of sodium benzoate, formic acid or potassium sorbate, the dosage levels to be used are as indicated above with the rule that it is not allowed to add more that 10% liquid in receptacles on the prepared fruits. Preservation by pasteurisation or "self-pasteurisation" will need as additional steps: a) boiling with a minimum water addition (maximum 10%); b) filling of receptacles; c) hermetic closing followed by d) pasteurisation or "self-pasteurisation".

Fruit Sugar Preserves Technology

As a overall rule of thumb, a sugar concentration of about 60% in finished or processed fruit products generally insures their preservation. Preservation is not only determined by the osmotic pressure of sugar solutions but also by the water activity values in the liquid phase, which can be lowered by sugar addition; and by evaporation down to 0.848 aw; this value however does not protect products from mould and osmophile yeast attack. Maximum saccharose concentration that can be achieved in the liquid phase of the product is 67.89%; however higher total sugar quantities (up to 70-72%) found in products are explained by an increased reducing sugar solubility resulting from saccharose inversion.

Jams

The preservation of fruit by jam making is a familiar process carried out on a small scale by housewives in many parts of the world. Factory jam making has become a highly complex operation, where strict quality control procedures are employed to ensure a uniform product, but the manufacturing operations employed are in essence the same as those employed in the house.

Fresh or pre-cooked fruit is boiled with a solution of cane or beet sugar until sufficient water has been evaporated to give a mixture which will set to a gel on cooling and which contains 32-34% water. Gel formation is dependent on the presence in the fruit of the carbohydrate pectin, which at a pH of 3.2—3.4 and in the presence of a high concentration of sugar, has the property of forming a viscous semi-solid.

During jam boiling, all microorganisms are destroyed within the product, and if it is filled hot into clean receptacles which are subsequently sealed, and then inverted so that the hot jam contacts the lid surface, spoilage by microorganisms will not take place during storage. About 30% of the vitamin C present in fresh fruit is destroyed during the jam-making process, but that which remains in the finished product is stable during storage.

The high moisture content of jam makes it susceptible to mould damage once the receptacle has been opened and exposed from some time to the air. No problems of microbiological spoilage are likely to arise in the canned product during storage.

Marmalade

This sugar preserve is defined as "semisolid or gel-like product prepared from fruit ingredients together with one or more sweetening ingredients and may contains suitable food acids and food pectins; the ingredients are concentrated by cooking to such a point that the TSS—Total Soluble Solids—of the finished marmalade is not below 65%".

Fruit Paste

Fruit paste is a product obtained in the same way as special non-gelified fruit marmalade but with a lower water content—about 25% TSS in fruit paste. Lowering water content could be achieved by continuing boiling of the product or by drying the product by natural or artificial drying. An example of paste without sugar is the sun dried apricot or prune paste.

Preparation of Jams, Jellies and Marmalade

— Boil the pulp or the juice (with water when necessary)

- Add the pectin
  - to the batch while stirring very vigorously
  - Pectin which has previously been mixed with 5 times its weight in sugar taken from the recipe)
- Boil for about 2 minutes to assure a complete dissolution
- Add the sugar while keeping the batch boiling
- Boil down quickly to desired Brix
- Add the acid (usually citric acid) and remove the froth
- Fill hot into the (previously cleaned) jars and close
- Invert the jars for three minutes to pasteurise the cover

Pineapple-papaya Jam

The fruit should be prepared as per previous instructions. For pineapples, the ends are removed and discarded; the cores and outer parts of the fruits are also removed. The fruit cylinders obtained are pulped through a special extractor (Fitzpatrick communiting machine) equipped with a 0.40-in screen sieve; the pulp thus obtained is used for making jam. The papaya are prepared by hand-peeling the fruit; the fruit is then halved and the seeds removed. It is then pulped in the communiting machine using a 0.40-in screen sieve. When ginger root is used as flavouring, it is peeled and macerated in a Kenwood blender to a very fine consistency.

A typical formula for a pineapple-papaya jam (50:50 ratio) with ginger flavouring is given as follows:

Pineapple pulp	25.0
Papaya pulp	25.0 pounds
Cane sugar	50.0
Apple pectin (150 grade)	6.0 ounces
Citric acid	6.4
Fresh ground ginger	7.5

Processing

The weighed fruit pulp is placed in a stainless steel steam-jacketed kettle and heated to about 110°F under constant stirring. When the product reaches this temperature, the heat is turned off. The pectin (mixed in about ten times its weight with some of the weighed sugar), is then mixed into the fruit pulp, stirring constantly in order to prevent the pectin from clotting.

When the pectin has dissolved, the remainder of the sugar is added and dissolved completely in the mixture. The heat is then turned on and the jam mixture is stirred constantly until it starts boiling vigorously. During the remainder of the cooking, the product is stirred occasionally. Near the finishing point (approximately 221° F), the citric acid and the ginger (if it is used) are also added.

Determination of the finishing point is done by removing samples at intervals, cooling, and reading the soluble solids by means of a refractometer equipped with a Brix scale. After the jam reaches the proper Total Soluble Solids content, the heat is turned off and the surface scum/foam is removed.

The jam then is quickly put into receptacles which have been cleaned and sterilised with boiling in water for 30 minutes. The filling operation is done rapidly in order to prevent the temperature of the jam from falling below 190° F. After filling, sterilised lids (boiled for 30 minutes in water) are placed on the receptacles and they are then sealed. After this operation the receptacles are inverted for about 3 minutes to insure that the lids are sterilised. The receptacles are then placed upright. At this stage it is not necessary to do any further processing, therefore the receptacles are cooled in running cold water until they reach a temperature slightly above room temperature. They are then dried in air and labelled.

Finished Products Evaluation

During production at medium / large scale, it is recommended that quality controls be performed during manufacturing. After ten weeks of storage at room temperature it is recommended that an examination of finished products be performed. The receptacles are opened and contents carefully emptied on to enamel trays without disturbing the formation of the jam. The empty cans are then inspected for signs of corrosion. Factors other than flavour include colour, appearance, syrup separation, firmness and spreading quality. For flavour, jam is tested on pieces of bread.

Samples are taken for measurement of pH (with a glass electrode pH meter) and Total Soluble Solids (with a refractometer equipped with a Brix scale). This evaluation enables to have a quality check during product shelf life and to obtain data needed for necessary improvements of future productions. For pineapple-papaya jam, products made with 30% pineapple and 70% papaya with added ginger has the highest score for flavour. The use of plain tin cans causes corrosion problems which is not the case when acid resistant lacquer cans are used

Technology of Fruit Juices

Fruit juices are products for direct consumption and are obtained by the extraction of cellular juice from fruit, this operation can be done by pressing or by diffusion. For the purpose of this chapter, the technology of fruit juice processing will cover two finished product categories:

— juices without pulp ("clarified" or "not clarified");

— juices with pulp ("nectars").

We will also define as:

— "natural juices" products obtained from one fruit; and

— "mixed juices" products obtained from the mix of two or three juices from different fruit species or by adding sugar.

Juices obtained by removal of a major part of their water content by vacuum evaporation or fractional freezing will be defined as "concentrated juices".

Technological Steps for Processing of Fruit Juices without Pulp

Fruit juices must be prepared from sound, mature fruit only. Soft fruit varieties such as grapes, tomatoes and peaches should only be transported in clean boxes which are free from mould and bits of rotten fruit.

Washing: fruit must be thoroughly washed. Generally, fruit will be submitted to a pre-washing before sorting and a washing step just after sorting.

Sorting: removal of partially or completely decayed fruit is the most important operation in the preparation of fruit for production of first quality fruit juices; sorting is carried out on moving inspection belts or sorting tables.

Crushing/Grinding/Disintegration Step is applied in different ways and depends on fruit types:

— Crushing for grapes and berries;

— Grinding for apples, pears;

— Disintegration for tomatoes, peaches, mangoes, apricots.

This processing step will need specific equipment which differs from one type of operation to another.

Enzyme Treatment of crushed fruit mass is applied to some fruits by adding 2-8% pectolitic enzymes at about 50° C for 30 minutes. This optional step has the following advantages: extraction yield will be improved, the juice colour is better fixed and finished product taste is improved. However, for fruit which is naturally rich in pectic substances, this treatment makes the resulting "exhausted" material useless for industrial pectin production.

Heating of crushed fruit mass before juice extraction is an optional step used for some fruit in order to facilitate pressing and colour fixing; at same time, protein coagulation takes place.

Pressing to extract juice.

Diffusion is an alternative step for juice extraction and can be carried out discontinuously or in batteries at water temperature of about 80-85 ° C.

Juice Clarifying can be performed by centrifugation or by enzyme treatment. Centrifugation achieves a separation of particles in suspension in the juice and can be considered as a pre-clarifying step. This operation is carried out in centrifugal separators with a speed of 6000 to 6500 RPM. Enzyme clarifying is based on pectic substance hydrolysis; this will decrease the juices' viscosity and facilitate their filtration.

The treatment is the addition of pectolitic enzyme preparations in a quantity of 0.5 to 2 g/l and will last 2 to 6 hours at room temperature, or less than 2 hours at 50° C, a temperature that must not be exceeded. The control of this operation is done by checking the decrease in juice viscosity. Sometimes, the enzyme clarifying is completed with the step called "sticking" by the addition of 5-8 g/hl of food grade gelatine which generates a flocculation of particles in suspension by the action of tannins.

Filtration of clarified juice can be carried out with kieselgur and bentonite as filtration additive in press-filters (equipment).

De-Tartarisation is applied only to raisin juice and is aimed to eliminate potassium bitartrate from solution. This step can be performed by the addition of 1% calcium lactate or 0.4% calcium carbonate. Pasteurisation of juice can be done for temporary preservation (pre-pasteurisation) and in this case this operation is carried out with continuous equipment (heat exchangers, etc.); warm juice is stored in drums or large size receptacles (20-30 kg). Pasteurisation conditions are at 75°C in continuous stream.

Pasteurisation of bottled juice is then carried out just before delivery to the market; this is performed in water baths at 75° C until the point where the juice reaches 68° C. In cases when the final pasteurisation is done without pre-pasteurisation and temporary storage, modern methods use a rapid pasteurisation followed by aseptic filling in receptacles. Rapid pasteurisation conditions are as follows: temperature about 80° C, over 10-60 sec., followed by cooling; all operations are carried out in continuous stream.

Preservation under CO_2 pressure may be done at a concentration of 1.5% CO_2 under a pressure of 7 kg/cm². At the distribution step, proceed at CO_2 decompression and the juice is then submitted to a sterilising filtration and aseptic filling in receptacles. Preservation by freezing is carried out at about -30° C, after a preliminary de-aeration; storage is at -15 to -20° C. Production of concentrated juices by evaporation is performed under vacuum (less than 100 mm Hg residual pressure) up to a concentration of 65-70% total sugar which assures preservation without further pasteurisation.

Modern evaporation installations recover flavours from juices which are then reincorporated in concentrated juices. Additional operations for juice manufacturing are the vacuum de-aeration and mixing with other fruit juices or with sugar. For the production of non clarified juices the centrifugation is the only specific step, enzyme clarifying and subsequent filtration being eliminated. The optimum sugar/acid ratio for the majority of fruit, mainly for pomaces, is 10/1 to 15/1. Fruit which is rich in carotenoids is only processed as juices with pulp ("nectars").

Technological Flow-sheet for Nectars

This process is divided at industrial scale in two categories of operations:

— processing for obtaining juices;

— juice conditioning for preservation.

Operations in the first category are differ according to the type of fruit which to be processed: Pomaces (apples, pears) are washed and sorted and then crushed in a colloid mill; fruit purée is then passed through a screw type heating equipment where direct steam is used as a source of heat. Warm fruit mass is treated in a pulper with a 2 mm screen and then through an extractor similar with the equipment used for tomato juice.

Stone fruits (apricots, peaches, cherries, etc.) after washing and sorting are submitted to steam in a continuous heater, then the warm fruit mass is passed through a pulper and then an extractor (as mentioned above). Berries (strawberry, wild berries, etc.) are washed, sorted and then crushed, preheated and then introduced in extractor. In order to avoid browning and undesirable taste modifications it is usual to add about 0.05% ascorbic acid.

Second category type of operations are similar for all fruit species: Partial elimination of cellulose is achieved with a continuous centrifugal separator; the resulting juice is then processed in order to adjust sugar and acid content for viscosity. Sugar (about 8-10%) is added as a syrup (in water or in the juice of same fruit obtained by pressure). Acidity is adjusted with citric or tartaric acid. The adjusted juice is then deaerated under vacuum at about 40° C. This step aims at avoiding oxidative reactions and vitamin C loss reduction.

An important subsequent step is an intensive homogenisation (under pressure at 150-180 A) in order to obtain particles with dimensions below 100. The homogenised juice obtained is then continuously pasteurised in plate heat exchanger equipment at a temperature of about 130° C, cooled down to about 90° C and aseptically packed in receptacles. The principal characteristics of fruit "nectars" are uniformity and stability of the content provided by the advanced disintegration of fruits.

Stability can be obtained by increasing product viscosity by adding pectin for fruit which is deficient in this component. In order to avoid "separation", intensive homogenisation is carried out as described above. Fruit "nectars" contain all the important components of the original fruit and to a large extent maintain their taste and flavour. The sugar/acidity (as citric acid) ratio is to a large extent determined by the type of fruit and the correction applied; for example, this ratio is 30 for apricots, 40 for peaches, 160 for pears, etc.

Banana and Plantain Processing Technologies

Traditional Processing

Products: Uses and Dietary Significance

Most of the world's bananas are eaten either raw, in the ripe state, or as a cooked vegetable, and only a very small proportion are processed in order to obtain a storable product. This is true both at a traditional village level with both dessert and cooking bananas and when considering the international trade in dessert bananas.

In general, preserved products do not contribute significantly to the diet; however, in some localised areas the products are important in periods when food are scarce.

Probably the most widespread and important product is flour preparation from unripe banana and plantains by sun-drying. In Uganda, dried slices known as "mutere" are prepared for storage from green bananas, the dried slices being either used directly for cooking or after grinding into a flour. "Mutere" is used chiefly as a famine reserve and does not feature largely in the diet under normal conditions.

In Gabon, plantains are sometimes made into dried slices which can be stored and used on long journeys, and plantains are used in Cameroon to prepare dried pieces which are stored and ground as needed into flour for use in cooking a paste known as "fufu". Dried green banana slices are also used in parts of South and Central America and West Indies for preparing flour.

The other nutritionally important product is beer which is a major product in Uganda, Rwanda and Burundi where green banana utilisation is particularly high.

Preservation Methods and Processes

Drying. Both ripe and unripe bananas and plantains are normally peeled and sliced before drying, although banana figs are sometimes prepared from whole ripe fruit. Sun drying is the most widespread technique where the climate is suitable but drying in ovens or over fires is also practiced. In west Africa, plantains are often soaked and sometimes parboiled before drying. The slices of unripe fruit are normally spread out on bamboo frameworks; or a cemented area; or on a mat; or on a swept-bare patch of earth; or on a roof; or sometimes on stones outcrops or sheets of corrugated iron.

Oven-drying of ripe bananas is practiced in Polynesia as a mean of preserving the fruits, which are then wrapped in leaves and bound tightly to store until needed. In East Africa a method has been reported that involves drying the peeled bananas on a frame placed over a fire for 24 hr before drying in the sun, to accelerate the process.

Product stability and storage problems

There is little experimental data on the storage life of the traditionally made banana and plantain products.

Potential for scaling up of traditional processes to industrial level

Many banana products are now produced on an industrial scale, including the traditional banana figs and flour, and the processing techniques are described below. One of the main problems encountered has been the susceptibility of banana products to flavour deterioration and discoloration and in the past many products reaching the market have been of poor quality.

A great deal of research has been directed to overcoming these problems, although however good the resultant products are they cannot compare in flavour and other characteristics with the fresh banana fruit. Indeed, an important constraint on the large-scale development of banana processing is the lack of demand for banana products since the fresh fruit is available throughout the year in most parts of the tropical world.

The production of beer from banana and plantains has not been scaled up to an industrial level, and while an important product in localised areas of tropical Africa, the market is rapidly declining in favour of European-type brews produced locally.

Industrial processing

Products and uses

The main commercial products made from bananas are canned or frozen purée, dried figs, banana powder, flour, flakes, chips (crisps), canned slices and jams.

Banana products can be divided roughly into two types - those for direct consumption, such as figs, and those for use in food manufacturing industry, for example purées and powder.

Banana figs, or fingers as they are sometimes known, are usually whole, peeled fruit carefully dried so as to retain their shape, although sometimes the fruit is sliced or halved to facilitate drying. Banana and plantain chips (crisps) are thinly sliced pieces of fruit fried in oil and eaten as a snack like potato chips (crisps).

The main use of canned slices is in tropical fruit salads. Banana flakes are used as a flavouring or in breakfast cereals. Banana purée find use mainly in the production of baby foods. Banana flour is said to be highly digestible and is used in baby and invalid foods, but can also be used in the preparation of bread and beverages. Banana powder is used chiefly in the baking industry for the preparation and fillings for cakes and biscuits and is also used for invalid and baby foods.

Processing technology

In general, to obtain a good-quality product from ripe-bananas the fruit is harvested green and ripened artificially under controlled conditions at the processing factory. After ripening, the banana hands are washed to remove dirt and any spray residues, and peeled. Peeling is almost always done by hand using stainless steel knives, although a mechanical peeler for ripe bananas has been developed, capable of peeling 450 Kg of fruit per hour.

The peeling of unripe bananas and plantains is facilitated by immersing the fruit in hot water. For example, immersion in water at 70-75 ° C for 5 min. has been suggested as an aid for peeling green bananas for flour production, while the peeling of green bananas for freezing has been facilitated by immersion in water at 93° C for 30 min.

Banana figs

Fully ripe fruits with a sugar content of about 19.5% are used and are treated with sulphurous acid after peeling, then dried as soon as possible afterwards. Various drying systems have been described using temperatures between 50 and 82° C for 10 to 24 hr to give a moisture content ranging from 8 to 18% and a yield of dried figs of 12 to 17% of the fresh banana on the stem.

One factory in Australia uses a solar heat collector on the roof to augment the heat used for drying bananas. Bananas can also be dried by osmotic dehydration, using a technique which involves drawing water from 1/4-in. thick banana by placing them in a sugar solution of 67 to 70 deg. Brix for 8 to 10 hr. followed by vacuum-drying at 65 to 70° C, at a vacuum of 10 mm Hg for 5 hr. The moisture content of the final products is 2.5% or less, much lower than that achieved by other methods.

Banana purée

Banana purée is obtained by pulping peeled, ripe bananas and then preserving the pulp by one of three methods: canning aseptically, acidification followed by normal canning, or quick-freezing.

The bulk of the world's purée is processed by the aseptic canning technique. Peeled, ripe fruits are conveyed to a pump which forces them through a plate with 1/4-in. holes, then onto a homogeniser, followed by a centrifugal de-aerator, and into a receiving tank with 29in. vacuum, where the removal of air helps prevent discoloration by oxidation.

The purée is then passed through a series of scraped surface heat exchangers where it is sterilised by steam, partially cooled, and finally brought to filling temperature. The sterilised purée is then packed aseptically into steam-sterilised cans which are closed in a steam atmosphere.

Banana slices

Several methods for canning of banana slices in syrup are used. Best-quality slices are obtained from fruit at an early stage of ripeness. The slices are processed in a syrup of 25 deg. Brix with pH about 4.2, and in some processes calcium chloride (0.2%) or calcium lactate (0.5%) are added as firming agents.

A method for producing an intermediate-moisture banana product for sale in flexible laminate pouches has been developed. Banana slices are blanched and equilibrated in a solution containing glycerol (42.5%), sucrose (14.85%), potassium sorbate (0.45%), and potassium metabisulphite (0.2%) at 90 deg. C for 3 min. to give a moisture content of 30.2%.

Banana powder

In the manufacture of banana powder, fully ripe banana pulp is converted into a paste by passing through a chopper followed by a colloid mill. A 1 or 2 % sodium metabisulphite solution is added to improve the colour of the final product. Spray- or drum-drying may be used, the latter being favoured as all the solids are recovered.

A typical spray dryer can produce 70 kg powder per hour to give yields of 8 to 11% of the fresh fruit, while drum-drying gives a final yield of about 13% of the fresh fruit. In the latter method the moisture content is reduced to 8 to 12 % and then further decreased to 2 % by drying in a tunnel or cabinet dryer at 60° C.

Banana flour

Production of flour has been carried out by peeling and slicing green fruit, exposure to sulphur dioxide gas, then drying in a counter-current tunnel dryer for 7 to 8 hr. with an inlet temperature of 75° C and outlet temperature of 45° C, to a moisture content of 8%, and finally milling.

Banana chips (crisps)

Typically, unripe peeled bananas are thinly sliced, immersed in a sodium or potassium metabisulphite solution, fried in hydrogenated oil at 180 to 200° C, and dusted with salt and an antioxidant. Alternatively, slices may be dried before frying and the antioxidant and salt added with the oil. Similar processes for producing plantain chips have been developed.

Banana beverages.

In a typical process, peeled ripe fruit is cut into pieces, blanched for 2 min. in steam, pulped and pectolytic enzyme added at a concentration of 2 g enzyme per 1 kg pulp, then held at 60 to 65° C and 2.7 to 5.5 pH for 30 min.

In a simpler method, lime is used to eliminate the pectin. Calcium oxide (0.5%) is added to the pulp and after standing for 15 min. this is neutralised giving a yield of up to 88% of a clear, attractive juice. In another process banana pulp is acidified, and steam-blanched in a 28-in Hg vacuum which ensures disintegration and enzyme inactivation. The pulp is then conveyed to a screw press, the resulting purée diluted in the ratio 1:3 with water, and the pH adjusted by further addition of citric acid to 4.2 to 4.3, which yields an attractive drink when this is centrifuged and sweetened.

Jam

A small amount of jam is made commercially by boiling equal quantities of fruit and sugar together with water and lemon juice, lime juice or citric acid, until setting point is reached.

Product stability and spoilage problems

All dried banana products are very hydroscopic and susceptible to flavour deterioration and discoloration, but this can be overcome to some extent by storing in moisture-proof containers and sulphiting the fruit before drying to inactivate the oxidases.

The dried products are also liable to attack by insects and moulds if not stored in dry conditions, although disinfestation after drying by heating for 1 hr to 80° C or by fumigation with methyl bromide ensures protection against attack. Banana powder is said to be stored for up to a year commercially and flakes have been stored in vacuum-sealed cans with no deterioration in moisture, colour or flavour for 12 months.

Banana chips tend to have a poor storage life and to become soft and rancid. However, chips treated with an antioxidant have been stored satisfactorily at room temperature in hermetically sealed containers up to 6 months with no development of rancidity.

Quality Control Methods

In general a good quality product is obtained if fruit is harvested at the correct stage of maturity and, where appropriate, ripened under controlled conditions. For example, in the case of banana figs, the fruit should be fully mature (sugar content of 19.5% or above) or the final product is liable to be tough and lacking in flavour. However, if over-ripe fruit is used, the figs tend to be sticky and dark in colour, so the fruit must be fully yellow but still firm.

For banana flour, which is prepared from unripe bananas, the fruit is harvested at three-quarters the full-ripe stage and is processed within 24 hr. prior to the onset of ripening. If less mature fruit is used, the flour tastes slightly astringent and bitter due to the tannin content. Bananas harvested between 85 and 95 days after the emergence of the inflorescence, with a pulp-to-peel ratio of about 1.7, were considered to be most suitable for the deep-fat frying.

Other criteria suggested for assessing maturity were beta-carotene and reducing sugar content, both of which increase with increasing maturity and pH which decreases as the fruit ripens, and these should be, respectively, about 2000 µg/100 g, less than 1.5% and 5.8 or above. Browning was found to occur if the sugar content was higher than 1.5%. The determination of crude fat in processed chips is also considered to be a necessary quality control measure.

It is important to remove all impurities prior to processing of products, and this is done by washing to remove dirt and spray residues and control on the processing line so that substandard fruit can be removed.

Preparation Methods for Fresh Bananas and Plantains

The main ways of preparing fresh bananas for consumption are boiling or steaming, roasting or baking and frying. Boiling followed by pounding into "fufu" is also widely adopted in certain areas of the tropics.

Boiling or steaming

Plantains and bananas are often prepared simply by boiling in water, either in their peel or after peeling, and either ripe or unripe; if unripe, the fruit is scraped thoroughly after peeling to remove all traces of fibrous material. The boiled fruit is eaten alone or more usually accompanied by a sauce. This preparation technique is widely used in West Africa.

Roasting or baking

Unpeeled or peeled fruit, either ripe or unripe, is roasted simply by placing in the ashes of a fire or in an oven. This method is widely used in West

Africa, East Africa and the South Pacific islands. For example, ripe plantains are placed unpeeled in an oven and when partly brown and tender, removed and peeled, then replaced in the oven and roasted evenly.

Frying

Ripe or unripe plantains or bananas are often peeled, sliced and cooked in oil, particularly in West Africa and in parts of South America and the West Indies. Similar products are also made in East Africa. Typically, ripe plantains are peeled, cut into slices or split lengthways, and fried in palm oil or with groundnut oil, the pieces being served either hot with a sauce or with fried eggs, or cold as a snack.

Pounding

Pounding is a process, used particularly in West Africa, for preparing most perishable staple food crops including plantains, cassava, yams and cocoyams to obtain a paste or dough known as "fufu" (also spelled "foofoo", "foutou", "foufou"). The plantains are peeled or boiled and peeled after boiling and pounded in a wooden mortar, the resulting paste normally being eaten with soup or a spiced sauce of meat and vegetables, but sometimes after wrapping in leaves and steaming.

Mango and Guava Processing Technologies

Mango Processing Technologies

Mangoes are processed at two stages of maturity. Green fruit is used to make chutney, pickles, curries and dehydrated products. The green fruit should be freshly picked from the tree. Fruit that is bruised, damaged, or that has prematurely fallen to the ground should not be used. Ripe mangoes are processed as canned and frozen slices, purée, juices, nectar and various dried products. Mangoes are processed into many other products for home use and by cottage industry.

The mango processing presents many problems as far as industrialization and market expansion is concerned. The trees are alternate bearing and the fruit has a short storage life; these factors make it difficult to process the crop in a continuous and regular way. The large number of varieties with their various attributes and deficiencies affects the quality and uniformity of processed products.

The lack of simple, reliable methods for determining the stage of maturity of varieties for processing also affects the quality of the finished products. Many of the processed products require peeled or peeled and sliced fruit. The lack of mechanised equipment for the peeling of ripe mangoes is a serious bottleneck for increasing the production of these products.

A significant problem in developing mechanised equipment is the large number of varieties available and their different sizes and shapes. The cost of processed mango products is also too expensive for the general population in the areas where most mangoes are grown. There is, however, a considerable export potential to developed countries but in these countries the processed mango products must compete with established processed fruits of high quality and relatively low cost.

Green mango processing

Pickles.

The optimum stage of maturity should be determined for each variety used to make pickles.

There are two classifications of pickles - salt pickles and oil pickles. They are processed from whole and sliced fruit with and without stones. Salt is used in most pickles.

The many kinds of pickles vary mainly in the proportions and kinds of spices used in their preparation. One basic recipe for the study of the preparation and storage of pickles in oil is as follows:

Mango pieces	250 g	Tumeric powder	2 to 4 g
Salt	60 g	Fenugreek seeds	2 to 4 g
Mustard powder	30 g	Bengal gram seeds	2 to 4 g
Chili powder	20 g	Gingelly oil	20 to 30 g

The ingredients are mixed together and filled into wide-mouthed bottles of 0.5 kg capacity. Three days later the contents are thoroughly mixed and refilled into the bottles. Extra oil is added to form a 1-2 cm layer over the pickles.

Chutney

The product is prepared from peeled, sliced or grated unripe or semi-ripe fruit by cooking the shredded fruit with salt over medium heat for 5 to 7 minutes, mixed and then sugar, spices and vinegar are added. Cook over moderate heat until the product resembles a thick purée, add remaining ingredients and simmer another 5 min. Cool and preserve in sterilised jars.

Spices usually include cumin seeds, ground cloves, cinnamon, chili powder, ginger and nutmeg. Other ingredients such as dried fruits, onions, garlic and nuts may be added.

Drying/dehydration. immature fruit is peeled and sliced for sun-drying. The dried mango slices can be powdered to make a product called amchoo. The use of blanching, sulphuring and mechanical dehydration gives a product with better colour, nutrition, storability and fewer microbiological problems.

Ripe mango processing

Purée.

Mangoes are processed into purée for re-manufacturing into products such as nectar, juice, squash, jam, jelly and dehydrated products. The purée can be preserved by chemical means, or frozen, or canned and stored in barrels. This allows a supply of raw materials during the remainder of the year when fresh mangoes are not available.

It also provides a more economical means of storage compared with the cost of storing the finished products, except for those which are dehydrated, and provides for more orderly processing during peak availability of fresh mangoes.

Mangoes can be processed into purée from whole or peeled fruit. Because of the time and cost of peeling, this step is best avoided but with some varieties it may be necessary to avoid off-flavours which may be present in the skin. The most common way of removing the skin is hand-peeling with knives but this is time-consuming and expensive. Steam and lyepeeling have been accomplished for some varieties.

Several methods have been devised to remove the pulp from the fresh ripe mangoes without hand-peeling. A simplified method is as follows: the whole mangoes were exposed to atmospheric steam for 2 to 2 1/2 min in a loosely covered chamber, then transferred to a stainless steel tank.

The steam-softened skins allowed the fruit to be pulped by a power stirrer fitted with a saw-toothed propeller blade mounted 12.7 to 15.2 cm below a regular propeller blade. The pulp is removed from the seeds by a continuous centrifuge designed for use in passion fruit extraction. The pulp material is then passed through a paddle pulper fitted with a 0.084 cm screen to remove fibre and small pieces of pulp.

Mango purée can be frozen, canned or stored in barrels for later processing. In all these cases, heating is necessary to preserve the quality of the mango purée. In one process, purée is pumped through a plate heat exchanger and heated to 90°C for 1 min and cooled to 35° C before being filled into 30 lb tins with polyethylene liners and frozen at -23.50 C.

In an other process, pulp is acidified to pH 3.5, pasteurized at 90°C, and hot-filled into 6 kg high-density bulk polyethylene containers that have been previously sterilised with boiling water. The containers are then sealed and cooled in water. This makes it possible to avoid the high cost of cans.

Wooden barrels may be used to store mango pulp in the manufacture of jams and squashes. The pulp is acidified with 0.5 to 1.0% citric acid, heated to boiling, cooled, and SO2 is added at a level of 1000 to 1500 ppm in the pulp. The pulp is then filled into barrels for future use.

Slices

Mango slices can be preserved by canning or freezing, and recent studies have shown the feasibility of pasteurized-refrigerated and dehydro-canned slices. The quality of the processed product in all of these procedures will be dependent upon selection of a suitable variety along with good processing procedures. Thermal process canning of mango slices in syrup is the most widely used preservation method.

Beverages

The commercial beverages are juice, nectar and squash. Mango nectar and juice contain mango purée, sugar, water and citric acid in various proportions depending on local taste, government standards of identity, pH control, and fruit composition of the variety used. Mango squash in addition to the above may contain SO2 or sodium benzoate as a preservative. Other food grade additives such as ascorbic acid, food colouring, or thickeners may be used in mango beverages.

A short description of finished products found in literature is as follows:

— *mango juice*: prepared by mixing equal quantities of pulp (purée) and water together and adjusting the total soluble solids (TSS) and acidity to taste (12 to 15% TSS and 0.4 to 0.5% acidity as citric acid);

— mango nectar containing 25% purée can be prepared using the following procedure.

Nectar components	Brix of purée 15°	17°	20°
Purée	100	100	100
Sugar	45	43	40
Water	255	257	260

Commercial processing conditions may require the use of a preservative.

The pH is adjusted to approximately 3.5 by adding citric acid as a 50% solution.

The time of heat processing will vary with filling temperature, can size and viscosity of the juice or nectar.

Mango squash may be prepared according to flow-sheet described below; the finished product may contain 25% juice, 45% TSS and 1.2 to 1.5% acidity and may be preserved with sulphur dioxide (350ppm) or sodium benzoate (1000 ppm) in glass bottles.

Mango squash simplified flow-sheet.

Ingredients		
Mango pulp	900	900
Sugar	900	1100
Citric acid	18	15
Water	900	900

Mangoes are washed, stored, peeled with stainless steel knives. The pulp is prepared by using a pulper with fine sieve (0.025-in); Sugar is mixed with water and citric acid = syrup; The pulp is added to the syrup and mixed well; The mixture is strained trough cloth; The squash is heated at 85° C and bottles are filled and closed.

For additional heat treatment bottles may need to be maintained at a product temperature of 80°C for 30 minutes if the product is to be processed without preservatives. The bottles are then left to cool in water and stored at room temperature.

Two negative points must be avoided: presence of air bubbles (which is a source of quick deterioration) and separation of squash solids (giving an undesirable appearance). The means to avoid these two phenomena are described in the fruit juices section.

The squash quality is evaluated on the basis of the following characteristics: pH, titrable acidity, soluble solids, ascorbic acid (by 2,6 dichlorophenol indophenol method), specific gravity.

Dried/dehydrated.

Ripe mangoes are dried in the form of pieces, powders, and flakes. Drying procedures such as sun-drying, tunnel dehydration, vacuum-drying, osmotic dehydration may be used. Packaged and stored properly, dried mango products are stable and nutritious.

One described process involves as pre-treatment dipping mango slices for 18 hr (ratio 1:1) in a solution containing 40° Brix sugar, 3000 ppm SO2, 0.2% ascorbic acid and 1% citric acid; this method is described as producing the best dehydrated product. Drying is described using an electric cabinet through flow dryer operated at 60° C. The product showed no browning after 1 year of storage.

Drum-drying of mango purée is described as an efficient, economical process for producing dried mango powder and flakes. Its major drawback is that the severity of heat preprocessing can produce undesirable cooked flavours and aromas in the dried product. The drum-dried products are also extremely hydroscopic and the use of in-package desiccant is recommended during storage.

Canning.

This preservation technology is described in various technological flow-sheets in this bulletin.

Guava Processing Technologies

Guava purée

Guava purée is used in the manufacture of guava nectar, various juice drink blends and in the preparation of guava jam. The washed sound fruit is first passed through a chopper or slicer to break up the fruit and this material is fed into a pulper. The pulper will remove the seeds and fibrous pieces of tissue and force the reminder of the product through a perforated stainless steel screen. The holes in the screen should be between 0.033 and 0.045 in. The machine should be fed at a constant rate to ensure efficient operation.

The puréed material coming from the pulper is next passed through a finisher. The finisher is equipped with a screen containing holes of approximately 0.020 in. The finisher will remove the stone cells from the fruit and provide the optimum consistency to the product.

Perhaps the best way to preserve the guava purée is by freezing and the material passing through the finisher can be packaged and frozen with no further treatment. It is not necessary to heat the product to inactivate enzymes or for other purposes. The material can be frozen in a number of types of cartons and cans; however, a fibre box with a plastic bag inside is commonly used and is probably the less expensive.

It is also possible to can and heat process the guava purée and this can be accomplished by heating the purée to 195° F in an open double bottom kettle, filling into cans, closing the cans, inverting the cans for a few seconds, followed by cooling. Cans should be cooled rapidly to approximately 100° F before they are cased and stacked into warehouses.

Guava juice and concentrate

Guava juice can be used in the manufacture of a clear guava jelly or in various drinks. A clear juice may be prepared from guava purée that is depectinised enzymatically. About 0.1% pectin-degrading enzyme is mixed into the purée at room temperature; heating of the product at approximately 120° F will greatly speed the action of the enzyme. After 1 hr. clear juice is separated from the red pulp by centrifuging or by pressing in a hydraulic juice press. A batch-type or continuous-flow centrifuge can be used on the depectinised purée with no further treatment.

The clear juice after centrifuge or after press (and subsequent filtration) can be preserved by freezing or by pasteurization in hermetically sealed cans.

For shipment to overseas markets it may be advantageous to concentrate either the purée or the juice.

Recent Trends in Fruits Processing

New products

The number and variety of fruit and vegetable products available to the consumer has increased substantially in recent years. The fruit and vegetable industry has undoubtedly benefited from the increased recognition and emphasis on the importance of these products in a healthy diet.

Traditional processing and preservation technologies such as heating, freezing and drying together with the more recent commercial introduction of chilling continue to provide the consumer with increased choice. This has been achieved by new heating (e.g. UHT, microwave, ohmic) and freezing (e.g. cryogenic) techniques combined with new packaging materials and technologies (e.g. aseptic, modified atmosphere packaging).

The overall trend in new fruit and vegetable products is "added value", thus providing increased convenience to the consumer by having much greater variety of ready prepared fruit and vegetable products. These may comprise complete meals or individual components. The suitability of products and packages for microwave re-heating has been an important factor with respect to added convenience.

A Fresh Look at Dried Fruit

New fruit varieties and advance in drying technologies are putting a fresh twist on dried fruit applications. Fruits that have been introduced to the drying process include cranberries, blueberries, cherries, apples, raspberries and strawberries - not to mention the traditional mainstays of raisins, dates, apricots, peaches, prunes and figs.

Perceived as a "value-added" ingredient, dried fruit adds flavour, colour, texture and diversity with little alteration to an existing formula. The growing interest in ethnic cuisines in U.S.A. and the change to a more healthy way of eating, has also moved dried fruit considerably closer to the mainstream.

Found primarily in the baking industry, dried fruit is coming into its own in various food products, including entrees, side dishes and condiments. Compotes, chutneys, rice and grain dishes, stuffings, sauces, breads, muffins, cookies, deserts, cereals and snacks are all food categories encompassing dried fruit.

Since some dried fruit is sugar infused (osmotic drying), food processors can decrease the amount of sugar in formula - this is especially the case in baked products. Processors are making adjustments in moisture content of the dried fruit so that a varied range is available for different applications. An added bonus is dried fruits' shelf stability (a shelf life of at least 12 months). Dried fruit is more widely available in different forms, including whole dried, cut, diced and powders.

Citric Acid and its Use in Fruit and Vegetable Processing

Citric acid may be considered as "Nature's acidulant".

It is found in the tissues of almost all plants and animals, as well as many yeasts and moulds.

Commercially citric acid is manufactured under controlled fermentation conditions that produce citric acid as a metabolic intermediate from naturally-occurring yeasts, moulds and nutrients. The recovery process of citric acid is through crystallization from aqueous solutions.

Citric acid is widely used in carbonated and still beverages, to impart a fresh-fruit "tanginess". Citric acid provides uniform acidity, and its light fruity character blends well and enhances fruit juices, resulting in improved palatability. The amount of citric acid used depends on the particular desired flavour.

Sodium citrate is often added to beverages to mellow the tart taste of high acid concentrations. It provides a cool, distinctive smooth taste and masks any bitter aftertaste of artificial sweeteners. In addition, it serves as a buffer to stabilise the pH at the desired level. The high water solubility of citric acid (181 g/100 ml) makes it an ideal additive for fountain fruit syrups and beverages concentrates as a flavour enhancer and microbial growth inhibitor (preferably at pH < 4.6).

In processed fruits and vegetables, citric acid performs the following functions:

a. It reduces heat-processing requirements by lowering pH: inhibition of microbial growth is a function of pH and heat treatment. Higher heat exposure and lower pH result in greater inhibition. Thus the use of citric acid to bring pH below 4.6 can reduce the heating requirements. In canned vegetables, citric acid usage is greatest in tomatoes, onions and pimentos. For tomato packs, the National Canners Association recommends a pH of 4.1 to 4.3. In general, 0. 1% citric acid will reduce the pH of canned tomatoes by 0.2 pH units.

b. Optimise flavour: citric acid is added to canned fruits to provide for adequate tartness. Recommended usage level is generally less then 0.15%.

c. Supplement antioxidant potential: citric acid is used in conjunction with antioxidants such as ascorbic and erythorbic acids, to inhibit colour and flavour deterioration caused by metal-catalysed enzymatic oxidation. Recommended usage levels are generally 0.1% to 0.3% with the antioxidant at 100 to 200 ppm.

d. Inactivate undesirable enzymes: oxidative browning in most fruits and vegetables is catalysed by the naturally present polyphenol oxidase. The enzymatic activity is strongly dependent on pH.

Addition of citric acid to reduce pH below 3 will result in inactivation of this enzyme and prevention of browning reactions.

Cherry and Apricot Oils are Safe for Food Use

The oils obtained by cold pressing the kernels of the cherry (Prunus cerasus) and the apricot (Prunus armeniaca) have been declared acceptable for food use by the UK Ministry of Agriculture, Fisheries & Food's Advisory Committee on Novel Foods and Processes (ACNFP) by June 1993.

In its assessments of the safety in use of the cherry and apricot kernel oils, the ACNFP consider specifications that included data on fatty acid composition, the presence of natural antioxidants and the content of cyanide, mycotoxins and heavy metals.

The Committee says that it gave particular consideration to the possible presence in the oils of the cyanogenic glucoside amygdalin, from which cyanide is released by enzymic action when the kernels of cherry and apricot are crushed. Amygdalin was found to be absent from the cherry and apricot kernel oils.

The oils are obtained by the mechanical mincing and cold pressing of kernels extracted from cleaned cherry or apricot stones. After filtering, the oils are stored and are to be sold in a raw, unrefined state. The cherry and apricot kernel oils are high unsaturated and are expected to be used as speciality oils for salad dressings, baking and shallow frying applications.

The Use of Fruit Juices in Confectionery Products

During the last decade, the concept of fruit juices has gained immensely on consumer popularity. The majority of new non-alcoholic and alcoholic fruit drink products were a combination of syrups, fruit juices and flavours.

The confectionery industry followed suit and new products incorporated fruit juices as part of their confectionery formulations and processes. Fruit juice concentrates of high solids are often used instead of normal or single-fold juices.

Juice concentrates are made of pure fruit juices. The process starts with pressing fruits and obtaining pure fruit juice; this is stabilised by heat treatment which inactivates enzymes and micro-organisms. The next processing step is concentration under vacuum up to 40-65° Brix or 4-7 fold. The concentrates are then blended for standardisation and stored.

These fruit juice concentrates are often further stabilised by the addition of sodium benzoate and potassium sorbate and are usually stored away from light and are refrigerated or frozen.

Depectinised fruit juices are also used to prevent foaming in confectionery processes and are essential for use in clear beverage products. Fruit juice concentrates which are depectinised, and have added preservatives are called stabilised, clarified, fruit juice concentrates.

Fruit juices are used in confectionery products in conjunction with natural and artificial flavours which provides intense flavour impact and are cost-effective for a confectionery product.

The traditional concern in using fruit juice concentrates in confectionery applications has been the effect of the natural acids on the finished product, particularly the formation of invert sugar during processing.

This is a logical concern since concentrates contain differing amounts and types of acids. For example: apple, cherry, strawberry and other berries contain primarily malic acid. Grapes mainly contain tartaric acid. Cranberry is high in quinic acid. Citrus fruits and pineapple contain differing amounts of citric acid. The concentrates, when used, are normally buffered to a pH of 5-7 with sodium hydroxide.

In formulating products with fruit juice concentrates, the solids of the concentrate are considered as mostly reducing sugars and a reduction in corn syrup is made to compensate for equivalent amount of reducing sugar being added in the concentrate.

The exact replacement can be determined by measuring the D.E. of the concentrate to be added. In formulations when small amounts of concentrate are used (less than 1%), no adjustment is made since the reducing sugar contribution of the concentrate is not significant.

Fruit juice concentrates can also be used to provide a source of natural colour, in particular red colour. Grape, raspberry, cherry, strawberry and cranberry concentrates in small amounts are very effective in colouring cream centres.

The inclusion of fruit juices in confectionery products is now left up to the imagination of the manufacturer. These products must, of course, hold up to the standards of flavour integrity, and product excellence, during the shelf-life of these products.

References

Fellows, P., 2000. *Food Processing Technology, Principles and Practice*, 2nd Edition, Woodhead, Cambridge.

FAO. 1969. Sun drying fruits and vegetables in Sudan. Jackson, T.H. and Mohammed, B.B. FAO Report. Rome: FAO.

Hanke, H. 1992. The use of fruit juices in confectionery products.In 46th P.M.C.A. Production Conference, 1992.

Pottter, N.N. 1984. *Food Science.* 3rd Edition. AVI Publishing Co. Westport, Conn.

Torregiani, D. 1993. Osmotic dehydration in fruit and vegetable processing. In *Food Research International.*

5

Processing of Dairy Milk Processing

Milk producing animals have been domesticated for thousands of years. Initially, they were part of the subsistence farming that nomads engaged in. As the community moved about the country, their animals accompanied them. Protecting and feeding the animals were a big part of the symbiotic relationship between the animals and the herders.

In the more recent past, people in agricultural societies owned dairy animals that they milked for domestic and local (village) consumption, a typical example of a cottage industry. The animals might serve multiple purposes (for example, as a draught animal for pulling a plough as a youngster, and at the end of its useful life as meat). In this case the animals were normally milked by hand and the herd size was quite small, so that all of the animals could be milked in less than an hour—about 10 per milker. These tasks were performed by a dairymaid (dairywoman) or dairyman. The word dairy harkens back to Middle English dayerie, deyerie, from deye (female servant or dairymaid) and further back to Old English dæge (kneader of bread).

With industrialisation and urbanisation, the supply of milk became a commercial industry, with specialised breeds of cattle being developed for dairy, as distinct from beef or draught animals. Initially, more people were employed as milkers, but it soon turned to mechanisation with machines designed to do the milking.

Historically, the milking and the processing took place close together in space and time: on a dairy farm. People milked the animals by hand; on farms where only small numbers are kept, hand-milking may still be practiced. Hand-milking is accomplished by grasping the teats (often pronounced tit or tits) in the hand and expressing milk either by squeezing the fingers progressively, from the udder end to the tip, or by squeezing the teat between thumb and index finger, then moving the hand downward from udder towards the end of the teat. The action of the hand or fingers is designed to close off the milk duct at the udder (upper) end and, by the movement of the fingers, close the duct progressively to the tip to express the trapped milk. Each half or quarter of the udder is emptied one milk-duct capacity at a time.

The stripping action is repeated, using both hands for speed. Both methods result in the milk that was trapped in the milk duct being squirted out the end into a bucket that is supported between the knees (or rests on the ground) of the milker, who usually sits on a low stool.

Traditionally the cow, or cows, would stand in the field or paddock while being milked. Young stock, heifers, would have to be trained to remain still to be milked. In many countries, the cows were tethered to a post and milked. The problem with this method is that it relies on quiet, tractable beasts, because the hind end of the cow is not restrained.

Types of Milk Products

Cream and Butter

Today, milk is separated by large machines in bulk into cream and skim milk. The cream is processed to produce various consumer products, depending on its thickness, its suitability for culinary uses and consumer demand, which differs from place to place and country to country.

Some cream is dried and powdered, some is condensed (by evaporation) mixed with varying amounts of sugar and canned. Most cream from New Zealand and Australian factories is made into butter. This is done by churning the cream until the fat globules coagulate and form a monolithic mass. This butter mass is washed and, sometimes, salted to improve keeping qualities. The residual buttermilk goes on to further processing. The butter is packaged (25 to 50 kg boxes) and chilled for storage and sale. At a later stage these packages are broken down into home-consumption sized packs. Butter sells for about US$3200 a tonne on the international market in 2007 (an unusual high).

Skim Milk

The product left after the cream is removed is called skim, or skimmed, milk. Reacting skim milk with rennet or with an acid makes casein curds from the milk solids in skim milk, with whey as a residual. To make a consumable liquid a portion of cream is returned to the skim milk to make *low fat milk* (semi-skimmed) for human consumption. By varying the amount of cream returned, producers can make a variety of low-fat milks to suit their local market. Other products, such as calcium, vitamin D, and flavouring, are also added to appeal to consumers.

Casein

Casein is the predominant phosphoprotein found in fresh milk. It has a very wide range of uses from being a filler for human foods, such as in ice cream, to the manufacture of products such as fabric, adhesives, and plastics.

Cheese

Cheese is another product made from milk. Whole milk is reacted to form curds that can be compressed, processed and stored to form cheese. In countries where milk is legally allowed to be processed without pasteurisation a wide range of cheeses can be made using the bacteria naturally in the milk. In most other countries, the range of cheeses is smaller and the use of artificial cheese curing is greater. Whey is also the byproduct of this process.

Cheese has historically been an important way of "storing" milk over the year, and carrying over its nutritional value between prosperous years and fallow ones. It is a food product that, with bread and beer, dates back to prehistory in Middle Eastern and European cultures, and like them is subject to innumerable variety and local specificity. Although nowhere near as big as the market for cow's milk cheese, a considerable amount of cheese is made commercially from other milks, especially goat and sheep.

Whey

In earlier times whey was considered to be a waste product and it was, mostly, fed to pigs as a convenient means of disposal. Beginning about 1950, and mostly since about 1980, lactose and many other products, mainly food additives, are made from both casein and cheese whey.

Yogurt

Yoghurt (or yogurt) making is a process similar to cheese making, only the process is arrested before the curd becomes very hard.

Milk Powders

Milk is also processed by various drying processes into powders. Whole milk and skim-milk powders for human and animal consumption and buttermilk (the residue from butter-making) powder is used for animal food. The main difference between production of powders for human or for animal consumption is in the protection of the process and the product from contamination. Some people drink milk reconstituted from powdered milk, because milk is about 88% water and it is much cheaper to transport the dried product. Dried skim milk powder is worth about US$5300 a tonne (mid-2007 prices) on the international market.

Other Milk Products

Kumis is produced commercially in Central Asia. Although it is traditionally made from mare's milk, modern industrial variants may use cow's milk instead.

Production of Milk

Milk and dairy products were in short supply and for the most part unavailable to those not living on or near the farm. Milk production was seasonal, creating periods of excess as well as deficiency in the family milk supply. Stabilisation of these production fluctuations by storage and/or further processing into butter, cheese, or other milk products was precluded by the lack of refrigeration.

Consequently, marketing of milk, butter, and cheese was limited to towns which could be reached by horse-drawn wagons. Over the years modern technology has rectified these problems and today a wide array of safe, wholesome dairy products are available to people throughout the developed world.

Production of quality milk is the concern of:

- consumers of dairy products
- retail distributors (super markets)
- milk and milk product processors

— dairy cooperatives
— state regulatory departments
— veterinarians, and
— dairymen.

From the list it's obvious that very few of us are left out. Whether we derive a living from the dairy industry through employment or otherwise, most of us are at the very least consumers of dairy products.

Composition and Nutritional Value of Milk

Milk is the lacteal secretion, practically free from colostrum, obtained by the complete milking of one or more healthy cows. Nearly 12% of the American household's total food expenditure is for dairy products. Milk and milk products alone provide 10% of the total available calories in the United States food supply, and in addition, represent one of the best natural sources of essential amino acids for human nutrition. These nutritional attributes of milk have long made it a mainstay particularly in the diet of growing children. There are estimated to be some 8 to 10,000 different types of milk products available thus making it an exceptionally versatile raw product.

Milk is composed of water, fat, protein, lactose and minerals (ash). The concentration of these components will vary between cows and breeds. Total milk solids refers specifically to fat, protein, lactose and minerals. This is to be differentiated from solids-not-fat milk (SNF), a frequently used term which describes the total solids content minus fat. SNF milk is known to most people as "skim milk". The nutritional as well as economic value of milk is directly associated with its solids content. The higher the solids content the better its nutritional value and the greater the milk product yields. For example, cheese yields are directly related to milk casein content.

Flavour and Odour

Consumer acceptance is greatly affected by flavour. There are several factors which may produce off-flavours and/or odours in milk. Some of the more common causes of flavor and odour problems are:

— Feed and weed flavours
— wild onion or garlic
— strong flavoured feedstuffs such as alfalfa silage

— Cow-barny flavours - which result when milk is obtained from unclean or poorly ventilated environments, improperly cleaned or sanitised milking equipment
— Rancid flavours - presence of free fatty acids (FFA)
— due to excessive agitation of milk during collection or transport
— breakdown of the milk fat component by proteolytic and lipolytic enzymes present in raw milk
— Malty flavours, high acid flavours
— bacterial contamination
— Oxidised flavours
— exposure of milk to sunlight
— contact of milk with oxidising agents such as rust, copper, and chlorine
— Foreign flavours
— fly sprays, medications, etc.

Objectives of Milk Processing

A multitude of events take place in the process of delivering milk from the farm to the dinner table and all are designed to provide the consumer with a wholesome, nutritious and safe product. The production of quality milk and milk products begins on the farm and continues through further handling, processing and distribution.

Milk processing has three primary objectives:

— destruction of human pathogens through pasteurisation
— keeping the quality of the product without significant loss of flavor, appearance, physical and nutritive properties, and
— selective control of organisms which may produce unsatisfactory products

Milk processing plant procedures seek to:

— prevent further bacterial contamination of raw materials
— reduce bacterial numbers in milk
— protect the finished product from recontamination through careful handling, proper packaging and storage

Pasteurisation is the means whereby raw milk is rendered safe for human consumption. It is the process of heating milk to a sufficient temperature for a sufficient length of time to make it free of pathogens, however, not totally free of bacteria.

Bacteria in Milk

As stated earlier, certain organisms are capable of surviving pasteurisation and/or refrigeration processes. These bacteria are an important concern because they reduce product shelf-life. Those of major significance are:

Thermoduric Bacteria

- common in raw milk
- they survive pasteurisation and include:
  - Enterococci
  - Micrococci
  - Brevibacterium
  - Lactobacilli

Psychrotropic Bacteria

- common dairy product contaminants
- these grow at refrigeration temperatures
- they do not survive pasteurisation
- can produce off-flavours

Spore-formers

- common contaminants
- survive pasteurisation
  - Clostridial spp.
  - Bacillus spp.

The primary source of these bacteria is the environment: air, dust, dirty equipment, operators, etc. Therefore, proper cleaning and sanitising procedures are necessary for quality control. Grade A milk quality standards allow a maximum of 100,000 bacteria/ml. in raw bulk milk. Chronic offenders of these limits risk losing their license to sell milk to the Grade A market. Most dairies are able to maintain bacteria counts between 5 to 10,000 per ml. When high counts become a problem it is generally due to one or more of the following:

- improper cleaning of milking equipment (the most common cause of high bacteria counts in milk)
- improper cooling of milk
- occasionally, a herd experiencing a high prevalence of infection due to Strep ag. or Staph sp.

Somatic Cell Counts

Somatic cell counts represent another important milk quality parameter. As discussed earlier, milk with high somatic cell concentrations reduces cheese yields due to the lowered casein content. In addition, high cell count milk generally contains increased amounts of proteolytic and lipolytic enzymes (lipase). These presence of these enzymes in milk increases the potential for off- flavours and odours. Somatic means body and thus a somatic cell is a body cell. There are three types of somatic cells typically found in milk: epithelial cells, macrophages, and polymorphonuclear leukocytes (PMN). Cell types found in milk obtained from non-infected glands are predominantly epithelial cells and macrophages.

Milk from infected glands, however, generally contains high concentrations of PMN's with little or no increase in other cell types. Consequently, somatic cell counts are an important indicator of udder health. Somatic cell counts are made available to dairymen from a variety of sources including milk quality laboratories operated by state and local regulatory departments, dairy cooperatives, DHIA-SCC program, and veterinary diagnostic laboratories. In general, cell counts from herd bulk milk consistently in excess of 500,000/ml are indicative of a high prevalence of mastitis in the herd.

Mastitis causes a shift in the composition of milk. In addition to lowered amounts of casein, lactose and fat levels are decreased particularly in milk with somatic cell counts in excess of 2 million. Because the bacterial quality and somatic cell content of raw milk are important to product shelf-life, flavor and yields (particularly cheese), milk processors strive to obtain the highest quality raw product possible from their producers.

Antibiotic Residues

Antibiotic residues pose a significant public health threat. Consequently, milk in Florida is routinely monitored by dairy cooperatives and the Florida Department of Agriculture and Consumer Services, Division of Dairy

Industry. The official test in current use is the Bacillus stearothermophilus disc assay. It is particularly sensitive for penicillin but can detect other inhibitors as well. The vast majority of antibiotic residues in milk occur by accident. Dairymen can avoid residue problems by:

— properly identifying treated cows
— informing milkers of the need to withhold and the method for withholding milk
— keeping an accurate record of dates and times of treatment
— following label directions and veterinarians' advice for withholding times
— having milk tested from suspect cows if uncertain about treatment or withholding time
— having tank milk tested when it is suspected of having milk containing antibiotic residue
— isolate purchased cows and test their milk for residue prior to their entry into the milking herd.

Proper Milking Procedures

Proper milking procedures are important for the prevention of mastitis and for insuring complete milk removal from the udder. Mastitis can decrease total milk production by 15 to 20%. To minimise loss and achieve maximum milk yield, a practical milking management scheme should be followed.

The term "milking management" includes care for the environment in which cows are housed or pastured. The dairy cow should have a clean, dry environment. This helps reduce the potential for mastitis and increases milking efficiency by reducing time and labour to clean udders before the milking process.

Movement of Cows

Movement of cows should be in a quiet, gentle manner. If cows are frightened or hurried, the milk letdown process may be disturbed. Therefore, rough handling of dairy cattle should be avoided.

Clinical Mastitis

Milking may begin with a check of all quarters for mastitis. It is acceptable to strip milk onto the floor in a milking parlour or flat barn. Any cows that show clinical mastitis should be examined and appropriate action taken. If fore milking is not done, visual checking for inflamed quarters is done by milkers and herd health people.

Dry Udders and Teats Preparation

The object of udder preparation is to ensure that clean, dry udders and teats are being milked. The federal government's pasteurised milk ordinance (PMO) states that a sanitizer must be applied before milking. This task may be accomplished by using an approved sanitizer injected in the floor-mounted cow washers or by using a hose and water with a sanitizer on the parlour. Single-service paper towels or washed and dried cloth towels may be used.

Predipping

Predipping with teat dip has become popular. The advantage may just be getting the water out of the milking barn so wet udders are not being milked. The procedure for predipping involves washing of teats with water and a sanitizer. The teats are then dried with an individual paper towel and dipped or sprayed with the sanitizer. A 30-second contact with sanitizer is needed to kill organisms. Then the sanitizer is wiped off of the teat with a paper towel.

The cows are milked and teats are dipped again with the same type of sanitizer to prevent chemical reactions that could cause irritation to teats. Predipping may be beneficial in reducing mastitis, but the actual dipping, dip contact time, and wiping with a towel increase the total milking time. If the dip is not wiped off, excessive chemical residues in milk may occur. If contact time is not sufficient, then it's a very expensive premilking regime.

Attach the Milking Unit

To attach the milking unit to the teats, apply the cluster allowing a minimum of air admission and adjust to prevent liner slip. Air entering the unit may cause the propulsion of mastitis organisms from one infected teat into a noninfected teat. This also may happen when one teat cup is removed before the others.

Machine stripping usually is not needed on dairy cows. Machine stripping should not take more than one minute and no air should be allowed to enter the teat cups while this is being done. A downward force applied to the cluster while massaging the udder with the other hand is all that is needed. Following milk-out, the machine should be removed only after the vacuum to the teat is shut off. This is accomplished most commonly by use of a vacuum shut off valve or milk hose clamp which prevents the backjetting of bacteria from one teat to another.

Backflushers

Backflushers have been developed to sanitize the liners and claws between milkings. Most units on the market have four or five cycles. The first cycle is a water rinse, followed by an iodine or similar sanitizer rinse, a clear water rinse, and positive air dry cycle. Research has demonstrated that backflushers do reduce the number of bacteria on the liners between cows, but do not reduce the number of bacteria on teats.

Backflushers also may stop the spread of contagious organisms, but this can also be accomplished at a much lower cost by teat dipping. There is no effect on environmental pathogens that are encountered between milkings. Backflushers may be effective in stopping the spread of contagious mastitis; however, there is limited research to support this view. Because of the high initial cost, the need for daily maintenance, and limited efficacy, backflushers are not routinely recommended.

Post-milking Teat Dipping

There is only one way to effectively stop the spread of mastitis in the dairy herd, and that is by applying teat dip to every quarter of every cow after every milking. Teat dips are used to remove milk residue left on the teat and kill organisms on the teat at the time of dipping. They also leave a residual film of sanitizer between milkings. Teat dips have been shown to effectively reduce mastitis caused by S. aureus and S. agalactia, the most common types of mastitis found in Florida.

Post-milking teat dipping is effective in eliminating environmental organisms E. coli and Strep. uberis on the teats after milking. These pathogens are found in the cow's surroundings; if there is udder-deep mud, the teat dip will be removed and a new infection may occur.

Types of Teat Dips

There are many effective teat dips, including iodine at 0.1%, 0.5%, and 1.0%. Also, although it is not labelled for teat dipping, hypochlorite at 4.0% with a sodium hydroxide content less than 0.05% was effective in field trials.

There are many more teat dips on the market that are effective in preventing new infections. Effective coverage of the teats is more important than the type of dip being used.

Dip or Spray

If contagious bacteria is present in your herd (Strep. ag., Strep. dysgalactiae, Staph. aureus, or mycoplasma), you must dip the whole teat to the base of the udder to stop the spread. Wand sprayers are acceptable for herds that have environmental mastitis, since teat colonisation is not a factor. Hand-held spray bottles are almost worthless in getting proper coverage of dip on the cow's teats, so they should not be used.

Requirements of a Milking System

Amount of Vacuum

The amount of vacuum used to operate the milking system is quite small, less than 2 CFM per unit. Extra CFMs are needed to compensate for vacuum losses that occur naturally in the system: head loss and resistance of pipes, elbows, etc. Thirty CFM ASME are needed in any system just to keep the system operating. Leaks account for a 10% loss (the difference between pump CFMs and system CFMs). Milk meters, unit slippage, air leakage used when applying the unit, and unit fall-off also consume varying amounts of vacuum depending on the skills of the operators.

Stated vacuum requirements vary greatly and are based on personal bias rather than research. The usual method is expressing CFMs per unit, which usually undersizes small systems and greatly oversizes large systems. Electric motors that run vacuum pumps are usually either 5, 7.5, 10, 15, or 20 HP and 1 HP motors will deliver 10 CFM ASME on an oil pump, or 7.5 CFM on a water pump. You must determine the pump size for your vacuum usage range.

For determining the size of vacuum pump needed, include 30 CFM for running the system, 10 percent loss due to leaks, and 3 CFM per unit for each unit (2 CFM to milk and 1 CFM added for milk meters and other losses). This should provide more than adequate vacuum for most systems. This does not include extra capacity for one half of the units open on the floor. A milking system should not be designed for poor milking procedures. Adequate milking vacuum level to prevent liner slip will reduce fall-offs and most automatic take offs will shut off automatically on fall-offs.

Milk Line Size

Research has indicated that milk lines have been oversized in the past. Oversized lines are expensive to purchase and expensive to clean, because they require more hot water and chemicals.

Pulsator Line Size

A 3-inch diameter plastic pulsator line is used for rigidity and ease of taping for pulsators. There is no need for a size larger than 3 inches.

Vacuum Controllers

Type: Diaphragm type are the most common and sensitive—Sentinel, Westfalia, Delaval Servo, DEC Servo, etc. Dead weight controllers should be avoided.

Location: In as clean an environment as possible and also easily accessible for cleaning. The vacuum controller should be located where the manufacturer recommends, usually between pump and the trap. If located on the balance or reserve tank, the diaphragm type should be double elbowed off the tank.

Maintenance: Clean as often as recommended or at least once a month. In systems with excessive amounts of pump capacity, cleaning must be carried out more often.

Pulsators

A pulsation rate of 40-60 pulsations per minute is adequate. Pulsation ratios of between 50:50 and 70:30 are adequate. A 70:30 ratio with 60 pulsations per minute gives the fastest milking. Alternate and simultaneous types are both acceptable. The choice of pulsation rates, ratios and type are more a matter of personal choice than based on any scientific data. Those commonly available from manufacturers work well. Pulsators should be cleaned monthly. Dust caps more often if dust, moisture and insects are common in the area.

Line Vacuum Level

Vacuum levels used are more based on choice than fact, as any vacuum between 12" Hg and 15" Hg may be adequate. Most high lines milk better at 15", low lines may work at 13.5" Hg. For maximum milking speed milk at 15" vacuum, 60 pulsations per minute, 70:30 ratio. But if "liner slip" or "fall off" are a problem, raising the vacuum to 15" usually helps. Vacuum levels above 15" Hg should be avoided. Line vacuum should be checked at the pump, the receiving jar, and by the controller.

Differences in vacuum levels between pump and receiver should be less than 0.5"Hg. Higher readings, indicating greater pressure drops, relate to decreased CFM at receiver. Greater pressure drops are influenced by small line sizes, excessive elbows, or unreasonably high air flows. Differences in vacuum levels between receiver and controller should be less than 0.2" Hg. Higher readings indicate higher vacuum differences which influence controller performance.

Role of Milking Machine

The role of the milking machine in causing mastitis has been debated for years. The most recent research has shown that a milking system properly installed, operated and maintained has little effect on mastitis. Research has demonstrated that cyclic vacuum fluctuation (that occurs normally in the milking cycle) and irregular vacuum fluctuation (that occurs when units fall off or are carelessly handled) occurring simultaneously will increase the new infection rate.

Line slip during milking will also increase the infection rate. Infection is increased because small droplets of milk are back-jetted against the teat end, with some organisms being forced into the teat. This is more of a problem with the way the unit is operated rather than with the milking machine itself. The most common way that the milking machine influences mastitis is that the system is not maintained properly and the milking machine may damage the teat end, thus increasing the chances of mastitis organisms entering the udder.

A vacuum level of above 15" Hg. will damage the teat end. This is caused by a malfunctioning or dirty vacuum controller. Malfunctioning vacuum controllers can also increase vacuum fluctuations. If the pulsators are dirty and no air is admitted, there will be a very short or no massage cycle which can cause damage to the teat end.

A recent survey in Florida indicated that 75% of the farms surveyed had pulsators malfunctioning and over 50% of the dairies surveyed had poor vacuum controller response. This survey indicates that these two important parts of the milking system are not being maintained. They must be for proper udder health! This fact sheet in conjunction with fact sheet DS-5, Advanced Milking Equipment Analysis Data Form, can give a complete analysis of the milking system. It also explains which parts of the system should be checked and how to do it. DS-5 also provides a record of this analysis for future reference.

Vacuum Pump

Evaluate every two months.

— Location - Pumps should be located in a well ventilated area, clean and dust free, as close to the parlour or milking barn as possible, but isolated enough to keep the noise level in the milking area to a minimum.

— Check belts for wear and tightness.

— Check rubber hoses for holes and hose clamp tightness.

— Check filters (if present) for cleanliness.

— CFM of pump should be checked with an orifice flow meter. The pump should be warm when checked. Do the check as close to the pump as possible, CFM is always checked at 15" vacuum, and recorded that way. Pump CFM's may also be checked at milking vacuum level if different than 15".

— Pump should function within 10% of factory rated capacity, if not, should be cleaned or rebuilt.

System Vacuum Level

Check several places in system. Resident gauge reading should also be checked: high lines 15", low lines 13.5" Hg.

System Check (Leaks)

Reconnect vacuum pump to the system, remove or shut off controller, shut off pulsators, shut off claws or crimp milk hoses and tape them. If possible remove top of receiver jar and insert the flow meter into receiver top adapter, start pump and record CFM's—leaks should not be more than 10% of pump capacity, but often are. If receiver jar is such that it can't be used, place flow meter in pipe where controller was removed. If leaks account for more than 10% of total capacity check pipe joint leaks, etc. Minimum system CFM's are 30 CFM (A.S.M.E.) + 1.5 CFM per unit.

Controller Responses

Reconnect system into milk mode, units not on, claw still off or hoses taped, controller in place and operating. Insert flow meter into receiver jar inlet or top, close flow meter, note vacuum level, let in air through flow meter in 10 CFM increments. Note vacuum: there should not be a vacuum change + or -0.5" vacuum up to 75% of capacity of system check CFM's. If controller response is poor, check for dirty filter(s) or dirty controller body. Most dead weight controllers will not pass this test, diaphragm type are most responsive.

Vacuum Controller Leakage (optional)

Remove or block off controller, place flow meter in receiver jar with system in milk mode (as in controller response check above) obtain CFM's at 0.5" Hg or 2 kpa below normal operating vacuum level, record CFM's. Reconnect vacuum controller with flow meter in same place, record CFM's at 0.5" Hg below vacuum level. The difference between the two should not be more than 10 CFM. Some diaphragm regulators use up to 10 CFM as a bypass procedure normally. Non-diaphragm regulators should not lose more than 1 CFM.

Pulsators

Should be inspected visually for cleanliness, air inlets should be clean, free of dust and spider webs. Pulsators used in flat barns should be checked for dented air inlets (where they are allowed to hit the floor). A graph should be made of each pulsator function while all other pulsators are operating on the system.

In flat barns both sides of the barn should be checked. Pulsation rate and ratio should be recorded and any abnormalities observed and recorded. If you do not make a graph a western dairy meter may be used. If nothing else is available, a vacuum gauge may be used to check if pulsator opens and closes—milking vacuum to 0" Hg.

Cluster

Pulsator hoses and short air hoses should be checked for leaks, holes, cracking, etc.

- Check claw vent: it should be open and clean if no liner vent is present, should be closed if liners are vented. If liner vents are used they should be clean and the claw vent sealed.
- A vacuum shut off should be present on the claw or on the milk hose before the claw.

— Liners or inflations should be free of holes and liners changed regularly; 1200 cow milkings for moulded liners and 600 cow milkings for stretch bore liners.

Automatic Take Offs: Sensor jars should be located at udder level or lower if possible. If floats are used they should be right side up. Older electronic sensing tubes should be checked for old collapsed rubber tubing.

Milk Line: Slope at least 1" per 10' (1 ½" per 10' is better) of line and should be looped, all inlets or nipples should be at the top of the line.

Vacuum Supply Line: If of P.V.C. it should be installed with enough supports to keep it from sagging and designed to be taken apart and washed.

Pulsator Lines: Should be supported to prevent low spots and have ports so it can be washed. Preferably looped and installed low enough that the pulsators can be reached without a ladder.

Stray Voltage: Anything above 0.5 volts A.C. should be avoided. A sensitive volt meter must be used and must not pick up D.C. voltage on the A.C. scale.

Places to Check

— Milk line to claw, one lead to the claw - the other to the milk line.

— Claw to the floor, one lead to the claw - the other to the floor.

— Bulk tank outlet to the milkhouse drain.

Wash System

— Hot water: temperature 160° preferable.

— Air injector must work for proper cleaning, a loud hiss at regular intervals usually means it works. Filter should be clean if present.

Dairy Processing Plant in Village

In developing countries, it is quite common to find milk production areas at great distances from areas of heavy demand for dairy products. These areas of demand are usually concentrated in the capitals and major cities. In an attempt to meet this demand in part, developing countries have tried to establish dairy industries near urban consumption centres. While there is overproduction in some parts of the country due to lack of outlets, dairy plants often work at only 20 percent of their capacity because of the long distances between the milk producing zones and the urban consumption centres where the plants are situated.

Costs of milk collection for plants located near the capital are very high due to fuel prices, poor road conditions (particularly during the rainy season), the cost of spare parts and the mileage required to be covered. For economic reasons, therefore, dairy plants have tried to restrict milk collection to nearby milk-producing zones, rounding out their daily output through the importation of dried milk and butter oil used for reconstituted milk. The cost of milk collection represents approximately 30 percent of the processing cost of the finished product (packaged pasteurized milk).

The traditional demand in developing countries is for fresh milk in the cities and fermented milk in the countryside. The dairy plant therefore primarily produces fresh and reconstituted milk, pasteurized, and packaged in plastic bags. Governments generally consider milk a basic commodity and consumer prices are strictly controlled. Plant profits from milk sales are usually quite low, forcing the management to market another, more remunerative product not subject to price control. A frequent choice is yoghurt, which is sold to wealthier people and to expatriates.

Besides the market for local milk products, there is also a market for imported products which, in order of importance, are dried milk for babies, tinned milk and cheeses. As a result, the problem in many developing countries is that while the small milk producer has no regular outlet for his milk, consumers pay high prices for imported dairy products.

First of all, a dairy processing plant should offer producers a guaranteed daily outlet for their milk supply. During the dry season, when domestic milk production is low, milk-pedlars are a common sight buying milk from producers who are far from urban consumption areas. However, during the rainy season, due to poor road conditions and increased milk production, this milk is not collected. If a milk producer were guaranteed a regular, fair income, it would provide the incentive for him to move from a position of self-sufficiency in milk for the family into one of selling, thus boosting milk production in his area.

Remote milk-producing areas are often good livestock production regions, but because of lack of commercial outlets they do not receive the necessary incentives for boosting milk production. The introduction of a dairy industry, modelled on European systems, into a developing country has only very rarely involved the producers themselves. And even when it has, usually only the big milk producers located near large towns have been involved.

The introduction of a processing plant or unit, at village level, concerns principally the local producers themselves. The size and simplicity of the unit should allow participation by small producers who are aware that self-help is a more immediate prospect than Government or bilateral assistance.

Village Milk Processing Unit

— by forming associations, the small producers, who live directly on the meagre resources brought to them by their cattle, can set up the unit. Many big herds belong to "entrepreneurs" in cities who entrust their stock to the care of herdsmen, paying them only the equivalent of one day's milking a week.

— by presenting a type of project which suits their needs—simple, practical and which allows them to work together within their own environment.

Milk-producing Area

— The site should be in a remote, inaccesible, traditional milk-producing area. Such areas, where communications are difficult and cattle-raising is a normal component of the family livelihood, are very common in Latin America, Africa, the Near East and Asia.

— An area should be chosen where milk collection from the capital is not possible.

— In an area where cattle, and milk in particular, should be the main regular source of cash for purchasing the family's domestic needs such as clothing, sugar, etc.

— In an area where there is plenty of water to allow hygienic processing of dairy products which requires an average of five litres of water for each litre of milk processed.

— In an area where the quanity of milk to be processed is available within a ten-mile radius since the time required for the transport of the milk should not exceed three hours. In tropical countries this is generally recognised as the upper limit beyond which milk cannot be pasteurized or heat-treated.

Locally-produced Dairy Products

— produces dairy products for which there is demand in the cities and in the villages;

— produces dairy products not supplied by dairy plants located near towns;

— produces dairy products which:

  — can easily compete with imported products;

  — do not require costly or sophisticated equipment;

  — require neither an expensive source of power nor a particularly complex infrastructure;

  — can be easily and cheaply moved without lowering the quality of the finished product;

  — can be sold in small unit sizes so as to reach a large number of consumers.

Locally-produced dairy products, particularly those such as cheeses, butter and fermented products, are frequently considered by the more affluent consumers in the capital cities of developing countries as second-rate compared to imported products.

In fact, locally processed milk products are often considered even by the producers themselves as merely by-products, surplus to their family needs. The project needs to demonstrate that with appropriate technology the finished products can be just as good as, and often better, than imported products considering the time required to clear imports through customs. At a minimal cost their presentation can also be as good.

Village dairy unit should be:

— as simple as possible;

— as clean as possible;

— as cheap as possible/

The village dairy unit is a tool:

— for a group, association or cooperative,

— for pooling resources so as to exploit their milk surpluses,

— to enable people to organise, and obtain the resources they need for improving their welfare.

General Model of Milk Processing Unit

Milk-producing Area

The unit should be located as centrally as possible within a given milk-producing area, near a source of water, or in a place where water is available. The site should be cool and well-ventilated. Sometimes not all these conditions can be met. The most important factor is availability of water. It should be remembered that on average five litres of water are required to process one litre of milk.

Building

Milk and dairy products are biologically active substances which are influenced by their environment. Cheese quality and reliability depend largely on the surroundings in which cheese is manufactured. An unused building can be purchased or leased and adapted for milk processing operations, or a new building can be constructed. It is not uncommon to find in some remote milk-producing areas, abandoned milk collection centres which may be suitable.

For a new building, the following factors should be taken into consideration:

— the walls should be built of local stone and the inner walls lined with a lime-cement mixture for easy cleaning;

— the cement floor should have a 2 to 3 percent slope for draining water used in cleaning;

— windows should be sufficient to provide adequate ventilation.

For village cheese manufacture, the second component of the building is the ripening cellar.

The most important features of a ripening cellar are maximum moisture (80 percent relative humidity) and low temperature (8 to 12°C). To achieve or approach these standards, a room partly below ground level is recommended. It should be about 2.5 m high. The floor of the room should be dug to a level of some 1.5 m below ground, with windows or openings made in the upper walls to lower the temperature of the cellar by a circulating draught, particularly important during the night.

Building size will of course depend on the quantity of milk received during the peak production period. An average quantity of milk which can be processed by a small-scale unit amounts to 100 to 500 litres per day. For these quantities the building area should be some 50 sq. m.

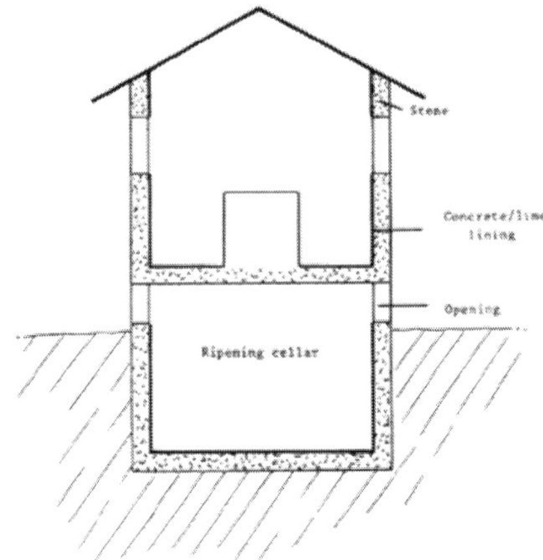

Figure 1. Section of the processing unit

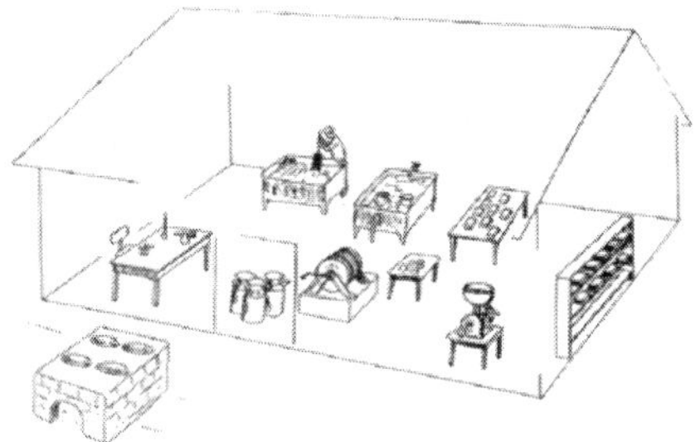

Figure 2. Diagram of building layout

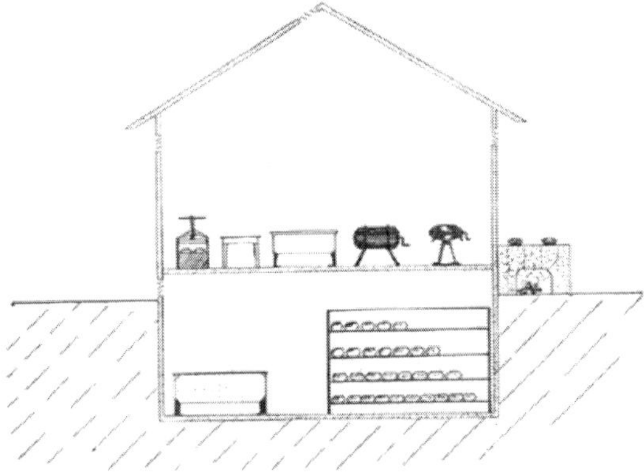

Figure 3. Sectional diagram of processing unit and equipment

Milk Processing Operations

As previously pointed out, milk processing operations will take place far from urban consumption centres. In these areas the quantities of milk hardly exceed an average of 500 litres per day. The products made must be able to withstand long periods of transportation, often under difficult climatic conditions.

Butter, cheese and processed cheese may require lengthy transportation, given the distance from consumer centres, whereas buttermilk and yoghurt can be marketed in the vicinity of the processing unit. The whey will be returned to the dairy farmers.

The main steps in obtaining the above products are:

— *Standardisation*: Standardisation is an operation producing milk with a constant butterfat content through partial, manual skimming. The operation makes it possible to standardise the composition of the finished product and to set aside part of the cream for butter.

— *Heat treatment*: Pathogenic germs in milk are destroyed by heating the milk to a minimum temperature of 63°C for 30 minutes.

— *Inoculation*: Due to heat treatment, which destroys a large number of lactic bacteria, cheese or yoghurt-making requires the addition of lactic bacteria to the milk. These bacteria are selected according to the type of finished product required.

— *Clotting*: Milk changes from a liquid to a solid state through the use of a coagulant: rennet.

— *Curd-Separation*: In cheese-making, the milk after coagulation is cut and separated into a liquid whey, and cheese curd.

— *Ripening*: This phase of cheese-making allows cheese texture to become homogeneous and the aroma to develop.

— *Churning*: In this operation, cream is churned to produce a semi-solid product which becomes butter.

— *Melting and emulsification*: Defective cheeses are melted and emulsified with salts to obtain a solid consistency and, after cooling forms processed cheese.

In addition to the eight steps mentioned above, milk collection, milk analyses and the marketing of the finished products should be equally regarded as important operations.

Equipment Needed

The equipment needed to run the dairy processing plant depends on several factors: how much milk is to be collected, how far and how scattered are the milk producers, what kind of product is to be produced? In the standard milk processing pattern, commencing with milk collection and ending with sale of dairy products, the following equipment would normally be required:

For Collection

- Plastic milking pails are often an improvement on the utensils commonly used.
- The producer should use small aluminium milk cans of 5-10 litre capacity for transporting the milk, while the collector should use 30-50 litre cans.
- If the milk needs to be collected, the following will be needed: the use of a bicycle, a 50 litre milk can, a graduated cylinder lacto-densitometer and a measuring pail.

For Processing

- *Reception*: The following equipment is needed for the reception of milk brought in by producers themselves and by the collector.
- *Storage*: A milk funnel and 50 litre milk cans.
- *Standardisation/Cream separation*: A manual cream separator has to be used to skim a portion of the milk received.

Heat Treatment

There are several possibilities for heat treatment, depending on the available power source. Under the least favourable circumstances, the sole available energy source is wood or peat. The best thing to use in this case is a "boiler/ water bath" as in the model below:

Figure 4. Cement-block boiler

More elaborate models can be built for wood or gas-fuelled heating (where bottled gas is available).

Figure 5. Wood-fuelled metal boiler **Figure 6. Gas-fuelled metal boiler**

For 100-500 litre quantities of milk, milk pasteurization with a plate pasteurizer is not recommended.

— *Cooling*: The milk is cooled with running water in a vat:

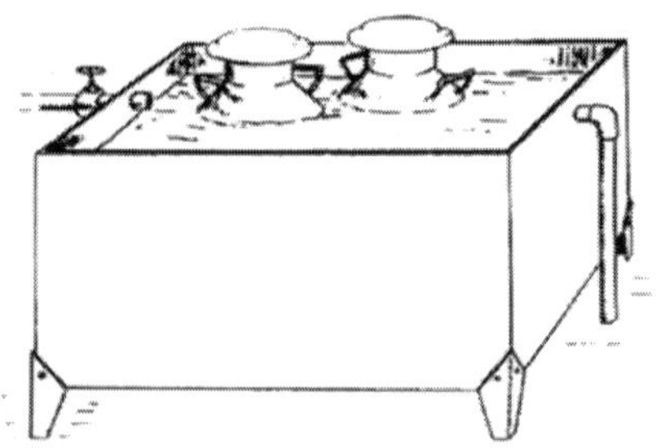

Figure 7. Simple cooling vat

or by water circulating in a jacketed vat.

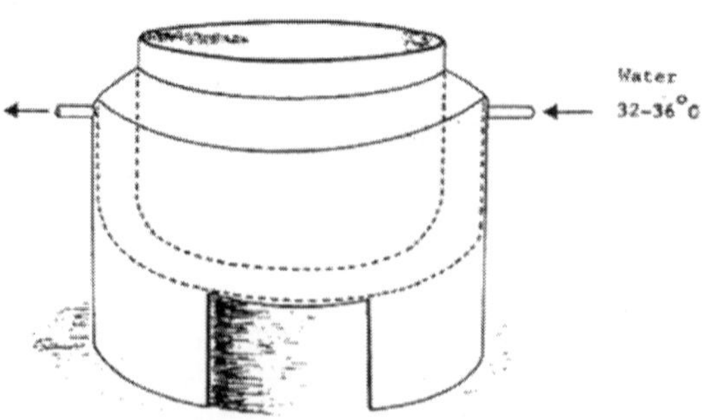

Figure 8. Jacketed cooling vat

— *Clotting*: The cheese vat can be of aluminium with a tap for draining the whey.

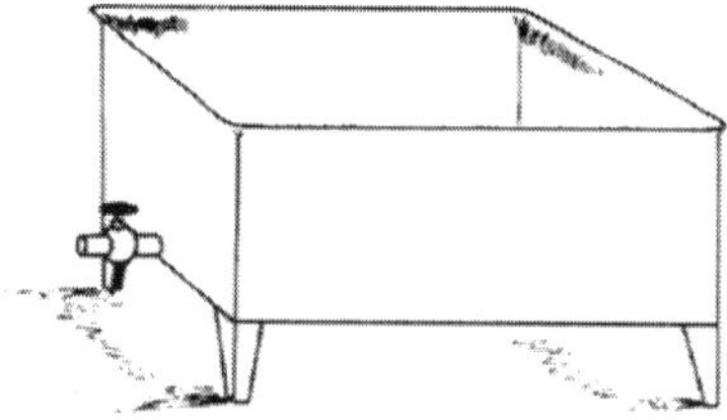

Figure 9. Simple milk clotting vat

A jacketed vat can also be used for milk clotting.

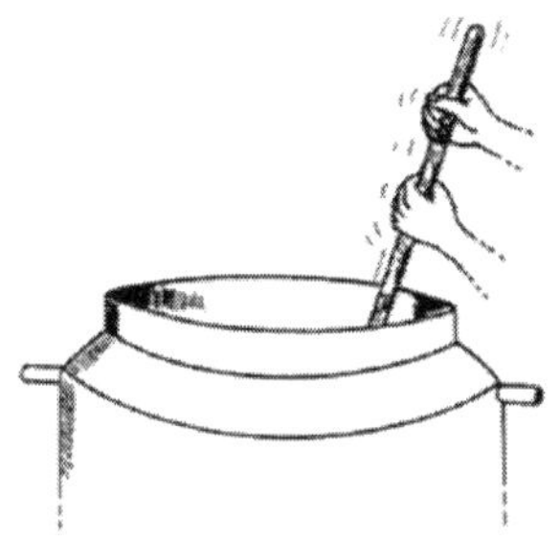

Figure 10. Jacketed clotting vat

There are other possible types of cheese vats, depending on the resources available to dairy producers. The following are a few examples:

— simple vat to process 50-100 litres of milk

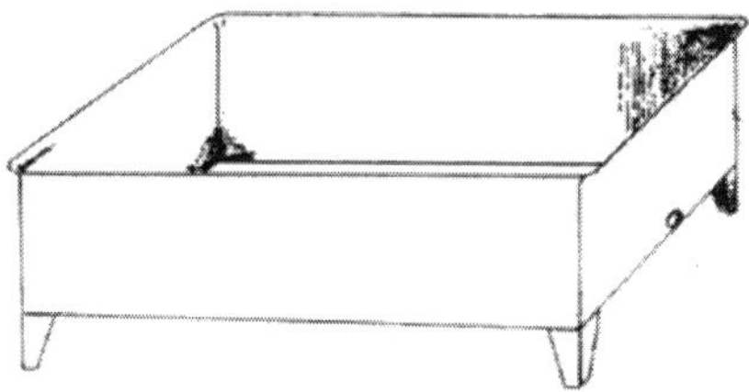

Figure 11. Low-walled clotting vat

— a more elaborate vat for processing 100-300 litres of milk: milk cooled by cold water circulating within its jacket:

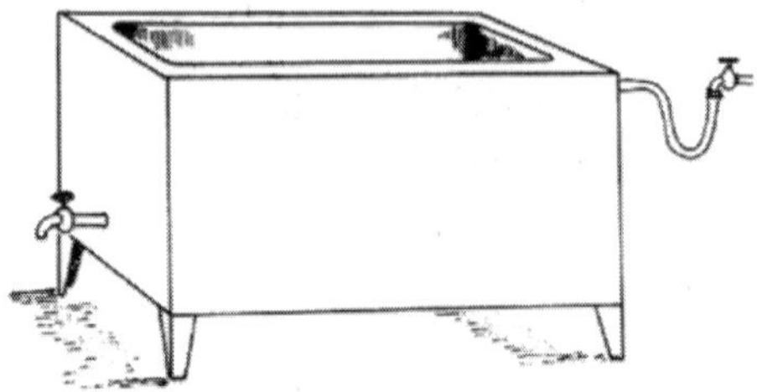

Figure 12. Clotting vat equipped for cooling

— for larger amounts of milk up to 500 litres, a vat equipped for heat treatment, cooling and clotting can be designed as shown in the following diagram:

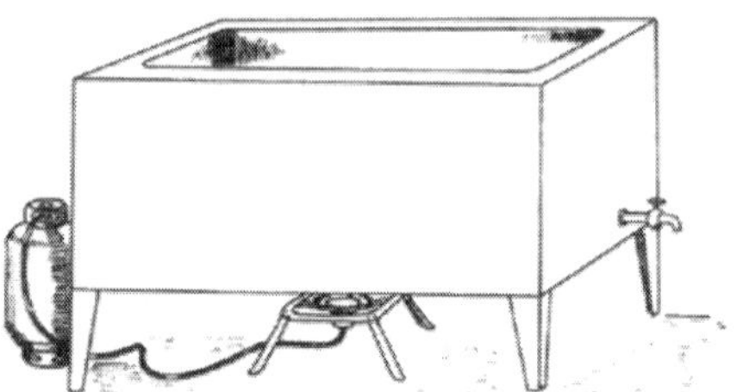

Figure 13. Multi-purpose vat

— *Draining*: The cheese is drained in moulds set on a slanting surface to drain the whey. The moulds, which give the cheeses their characteristic shapes, vary greatly in size and form.

The simplest way to make cheese moulds is to cut a plastic pipe generally used for drainage into 10 cm sections.

Figure 14. Plastic pipes

Bases and lids are of wooden discs slightly smaller in diameter than the cylinders.

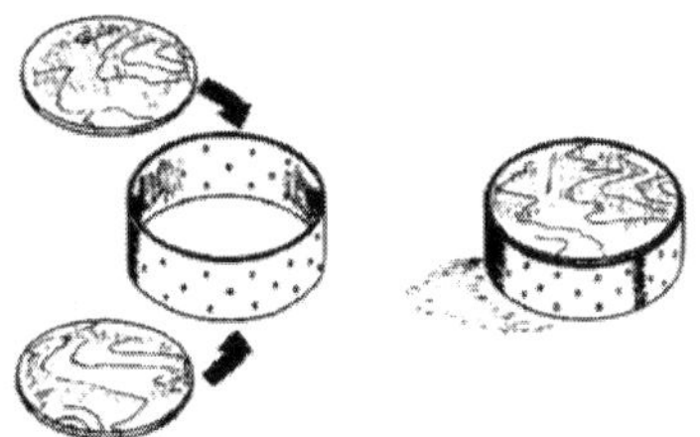

Figure 15. Moulds with bases and Lids

Cheese moulds may also be made of wood, their shape varying according to the type of cheese made.

The cheese-draining table, slanted forwards to facilitate draining of the whey, is made of wood.

Figure 16. Wooden cheese-draining table

— *Pressing*: Different kinds of cheese (curd) presses can be made. The simplest press is made by placing weights or cement block on the moulded cheese.

— *Ripening*: Wooden shelves must be assembled for the cheese-ripening cellar.

Figure 17. Ripening shelves

For Marketing

Marketing, as previously mentioned, is a very important aspect of overall operations. Quality of presentation is important. Wooden boxes made to accommodate the specific cheese shapes are used to transport the cheese to consumer areas.

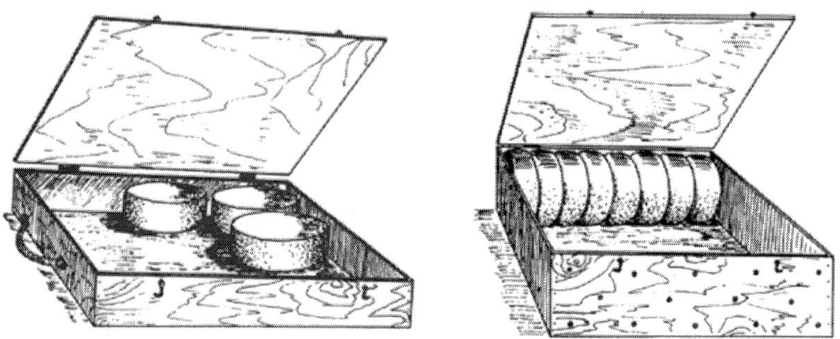

Figure 18. Wooden cheese boxes

Additional Equipment

An important component in a dairy processing unit, is the laboratory.

Laboratory equipment should include the following:

— To measure milk density:

 — Two or three lactodensitometers with glass cylinders

— To test milk acidity:

 — A Dornic acidimeter and the accessories shown below.

Salut acidimeter used for selective testing of milk acidity immediately after reception of farmers' milk:

— Fat content of milk

— Preparation of cultures

To prepare mesophilic cultures for cheese-making or thermophilic cultures for yoghurt-making, a strain of starter culture is initially required. Culturing and sub-culturing is done in individual 5, 10 and 15 litre containers.

Organisation and Operation of the Milk Processing Unit

A village milk processing unit usually involves a group of milk producers living within a given area near the unit. It is therefore reasonable that the majority of the producers will deliver their own milk in the morning direct to the processing unit. The milk from the evening milking is retained by the family for home consumption.

For producers who live further away or small farmers who do not wish to make the trip with a small quantity of milk, collection can be arranged within a radius of no more than 10 kilometres. A milk collector on a bicycle picks up the milk at collection points, which are simple shelters within which the milk can be kept in the shade. At the milk collection point, the milk collector tests milk density with the lactodensitometer, tests milk acidity, and measures milk volume using a measuring can. The milk is then filtered and poured into the milk can.

The entire milk collection round should take no longer than two hours. The milk collector has a little notebook in which he enters the volume and density of the milk delivered by the producer. Payments for the milk are made twice a month. The person responsible for milk collection may either be paid by the village cheese unit or by the producers themselves. For purposes of simplification, the second choice is preferable.

Reception of the Milk

Reception of the milk delivered by the farmers themselves or by the collector should take place in the very early morning. The producers' milk is weighed, density checked, the milk filtered, and then poured into milk pails.

Milk density is measured as shown in the diagram below:

— The producer's milk sample is poured slowly into a test-tube to avoid foaming;

— The lactodensitometer is introduced into the test-tube and, once the level is steady, the number is read:

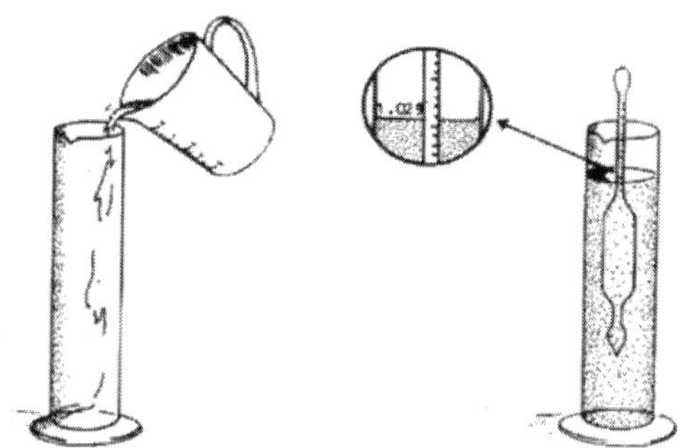

Figure 19. Interpretation of readings

For cow's milk an approximate estimation of density can be read as follows:

Lactodensitometer reading	Result
1.028 to 1.033	Normal milk
less than 1.028	Diluted milk
1.033 to 1.037	Skimmed milk

Standardisation of Milk

When milk from all the producers has been poured into the milk cans, a sample is taken from each can and mixed together to obtain an average sample. This sample is tested for milk acidity and fat content.

Milk Acidity

— Put 10 ml of the milk sample into a glass.

— Add 3 to 4 drops of phenolphthalein.

— With the dropper, add the NaOH $\frac{N}{9}$ solution drop-by-drop into the glass until a stable pink colour results.

— Read on the graduated column the number of ml used. This gives the acidity of the milk in Dornic degrees.

$$15\ ml\ of\ NaOH\ \frac{N}{9} = 15\ Dornic\ Degrees$$

Fat Content

1. Put 10 ml of sulphuric acid in the butyrometer.
2. Add 11 ml of milk from the average sample.
3. Add 1 ml of amyl alcohol.
4. Shake the butyrometer to dissolve the milk elements.
5. Put the butyrometers in the centrifuge. Centrifuging should continue five minutes.
6. Next plunge the butyrometers vertically, cork down, into a water bath, temperature 65°-70° C, and leave them for five minutes.
7. The butyrometer, cork down, should be perfectly vertical when taking the reading at eye level. Read the graduation mark at the base of the meniscus, i.e. at the base of the curved upper surface of the fat column.

In the example that follows, the degree reading is 3.6. The milk fat content then is 3.6 percent, or 36 g of fat per litre of milk.

Heat-treatment of Milk

Heat-treatment of milk is a most important factor in the quality of the finished product. After standardisation, milk must be heat-treated. This means bringing it to a minimum temperature of 63°C for 30 minutes. The various equipment suggested for heat-treatment can bring the temperature up to 63°C from between 40 to 60 minutes. The milk must be constantly stirred to maintain a homogeneous temperature throughout the heating and heat-treatment stages. The milk stirrer and the thermometer are important items of equipment.

Cooling Milk

When the milk has been kept at a temperature of 63°C for 30 minutes, it is then cooled down to bring it to a suitable temperature for cheese-making (approximately 35°C). The milk is cooled either by immersing the milk cans in a tank with cold running water, or by running cold water through the double sides of the cheese vat. During the cooling, as during the heating period, the milk must be stirred constantly.

Bearing in mind heat losses, cooling water should be shut off before the desired milk temperature is reached. For instance, if the milk temperature for cheese-making is 32°C, cooling water must be shut off when the milk temperature has reached 35°C.

Processing Aids

Preparation of Starter Cultures

One person only must be responsible for the preparation of starter cultures.

The simple method for preparing starter cultures is illustrated.

- The preparation of the starter cultures requires suitable strains of bacteria and a milk of good bacteriological quality.
- The starter culture used should be a commercial lyophilized one. These strains keep relatively well and it is, therefore, advisable to maintain a three-months' supply.

Following the heat-treatment of milk for cheese-making, a certain amount of milk for the preparation of the starter culture is retained in the milk can or in the jacketed vat in order to maintain the heat-treatment of 63°C for an additional 15-30 minutes. The milk is then poured into one-litre bottles and into a 5-litre container and covered with a clean cloth.

Imported Lyophilized Starter Culture

The contents of the vial are poured into a bottle of milk and shaken well to mix the powder and milk together.

Mother Culture

In order to avoid contamination by air, both the bottle containing the mother culture and the 5-litre container of starter culture are placed in a small wooden cupboard.

Starter Culture for Cheese-making

The mother culture to be used for cheese-making (mesophilic culture) is placed in the culture cupboard. The incubation temperature should be 20 - 22°C for 15-16 hours.

The acidity of the mother culture will then be 80 to 90°D. Use the mother culture to inoculate, at 2 percent, a second mother culture and 5 litres of milk for cheese-making. If the starter cultures are prepared with great care, one strain of commercial culture may be used for one to two months by means of successive subculturings.

Yoghurt-making Cultures

- Starter cultures for yoghurt-making are thermophilic bacteria. They must be cultured therefore at a temperature of 40 - 45°C for three to four hours.
- After heat treatment of the milk to which the yoghurt bacteria are to be added, the milk is cooled to 45°C and then poured into 1 litre bottles. These are then set into a water tank at 45°C to stabilise the bottled milk temperature.
- The lyophilized commercial strain for yoghurt-making is then put into the bottles.
- The bottles of milk are left in the water bath at 45°C for three to four hours.

Preparation of Rennet

Rennet, like starter cultures, may be imported. However, unlike commercial starter strains, rennet can be made locally. Since laboratory equipment is needed for making rennet, it should therefore be undertaken with the advice of a laboratory in the capital city—that for example of a university.

Obtaining Abomasa

Abomasa should be obtained from preferably unweaned calves. Assuming that the annual amount of calf rennet to be obtained is some 120 litres of liquid rennet at a strength of 1/10 000 and a yield of approximately 2 abosama for 1 litre of rennet, 240 abosama will be required per annum.

Preparation of Abosama

- The abosama should be washed and the fat and veins removed.
- The abosama are then inflated with air to avoid the two sides touching. They are then ball-shaped, the neck and the base being tied with string.
- The inflated abosama are hung in a dry, well-ventilated area. Drying should be complete after approximately one month of storage. At this stage, the flattened abosama can be kept in a dry place for a long period without signs of deterioration (about one year).

Soaking

When required for use, the abosamum should be sliced into thin strips, 5 mm wide. In an-easy-to-clean basin of plastic or stainless steel, soak the strips of abosamum in a 10 percent sodium chloride (salt) and 1 percent sodium benzoate solution. To produce a three months' batch of 30 litres of rennet, 60 abosama should be soaked in 40 litres of this brine solution.

The pH of the solution is adjusted to 4.3 with benzoic acid, after soaking for 24 hours at a temperature of from 20 to 25°C. Pour the liquid off into a container. In the same soaking basin and keeping the same abosama strips, the brine solution may be renewed four to five times before the properties of the abosama strips are completely exhausted. Each extract obtained in this fashion is treated as described above and the first and most concentrated extract is used to standardise the strength of the rennet.

Treatment of the Liquid

- To eliminate the mucilage in suspension in the extract, the solution is reacidified with hydrochloric acid to a pH of 4.8, and allowed to settle for two hours.

— The pH of the extract is raised with disodium phosphate to pH 5.5 - 5.6.

— Vigorous stirring must accompany these two actions. The liquid is then filtered over Watmann paper. Filtration may take a long time and probably will require distribution of the extract over several filters.

— The crude rennet extract obtained is generally a golden-yellow colour.

Determination of Strength

Definition: Rennet strength is the number of volumes of coagulated milk clotted by one volume of rennet in 40 minutes at 35°C. If "v" equals one volume of rennet, and "V" one volume of milk and measuring the clotting time in seconds, the calculation is:

$$S = \frac{2400\,V}{Tv}$$

In practice, liquid rennet strength should be 1/10 000 (1 litre of rennet clots 10 000 litres of milk at 35°C in 40 minutes).

Method

Put 500 ml of fresh milk in an Erlenmeyer flask and plunge it in a water bath at 35°C.

Remove 1 ml of the rennet to be standardised and dilute it in 10 ml of water.

When the milk in the Erlenmeyer flask reaches a constant temperature of 35°C, pour 10 ml of diluted rennet, stirring constantly, and start the timer. Keeping the Erlenmeyer flask in the water bath, slant it while rotating it gently so that a film of milk is formed on the sides of the flask. When the liquid begins to floculate, stop the timer.

Rennet strength: $S = \frac{2400 \times 500\ cc}{T \times 1cc}$

If flocculation time is 60 seconds, for example, rennet strength will be:

$$S = \frac{2400 \times 500}{60 \times 1}$$

$$S = 20\ 000$$

Standardisation

The strength of the four or five different batches, obtained by successive extractions of the rennet from the abosama is determined, thus enabling the rennet to be readjusted to a strength of 1/10 000. For example, if 30 litres at a strength of 1/20 000 have been obtained, and the extract from the four preceding rennet mixtures gave a solution at a strength of 1/5 000, the volume of rennet at strength 1/5 000 required to prepare standard rennet at a strength of 1/10 000 is determined in the following way:

$$30 \text{ litres} \times 20\,000 \text{ units} + \text{“Y” litres} \times 5\,000 \text{ units} = (30 + Y) \text{ litres} \times 10\,000 \text{ units}$$

$$600\,000 \text{ units} + 5\,000 \quad Y = 300\,000 + 10\,000\ Y$$

$$Y = 60 \text{ litres.}$$

After mixing the two extracts, 90 litres of rennet will be obtained at a strength of 1/10 000.

Maintenance-storage

Rennet must be stored in opaque glass bottles or in dark (blue or black-tinted) plastic containers and placed in cold storage at a temperature of 5 to 7°C. Under these storage conditions, the rennet will remain active for three months.

Renneting the Milk

It is recommended that coagulation tests be made first on small volumes of the cheese-milk. This is to redefine the amounts of rennet to use in order to obtain the same coagulation time as obtained previously with powdered rennet of 1/100 000 strength.

Technology for the Manufacture of Diary Products

Cheese

The following cheese-making operations are for making a firm-bodied cheese (Saint Paulin or Gouda type), a cheese most often produced in the developing countries and one which is relatively easily made to standard. The various steps of the cheese-making process should be adapted to climate, equipment available and consumer preferences.

The producers' milk is first filtered and then poured into the one-step "pasteurization/cheese-making vat". After filtering, a portion of the milk is passed through the separator to remove the required amount of cream. The milk used for cheese-making is standardised to a level of fat content of some 26 g/l. The separator can be a source of contamination, and it is therefore preferable to standardise before pasteurization rather than after.

The milk is pasteurized at low temperature in the double-walled vat at some 63°C for about 30 minutes. The source of heat can be gas or electricity or even fuelwood. The milk is then cooled to a temperature of 32-35°C by circulating cold water through the jacket of the vat. The lactic acid bacteria should be added at a rate of 1 to 2 litres per 100 litres of milk 15 to 20 minutes before renneting. The rennet (at a strength of 1/10 000) is added at a rate of 20 - 25 ml of rennet per 100 litres of milk. This is the time to add, if necessary, calcium chloride (5-50 g per 100 litres of milk).

Following this, flocculation time is from 10 to 15 minutes, and total clotting time is from 15 to 40 minutes. The curd is cut into regular grain-sized pieces. The first stirring should be carried out with both the curd grains and the whey, and should last from 5 to 10 minutes. Drawing off the whey (lactose removal) consists of extraction of a portion of the whey (from 20 to 60 percent), followed by the addition of an equal amount of water at a temperature of 30 to 35°C.

Potassium nitrate may be added at this stage of the cheese-making process. The second stirring, with moderate shaking of the curd grains in the diluted whey, lasts 10 to 20 minutes. The curd and some whey are then put into cloth-covered moulds. These moulds may be made of wood, stainless steel or plastic. Mechanical pressing lasts from 1 to 6 hours. During this operation, the cheese mould will be turned over 2 to 4 times. The cheese is salted in saturated brine at a temperature of 10 to 14°C. Brining time varies according to the volume of cheese.

A Saint-Paulin cheese weighing from 1.5 to 2 kg and with a diameter of 20 cm, is salted for about 8 hours. Working out the curd lasts 2 to 3 days, at a temperature of 10 to 12°C and a humidity rate of 80 to 85 percent. The cheese is placed on wooden shelves to ripen for at least 15 days at a temperature of 10 to 16°C with a humidity rate of 90 to 95 percent. Prior to sale, the cheese can be covered with a protective film of wax. If the cheese processing unit produces very large cheeses, they should be cut into 100 - 200 g slices prior to sale, and wrapped in greaseproof paper.

Cream and Butter

When standardised milk is used to make cheese, there will inevitably be excess fat in the form of cream. Often it is advantageous for the cheese unit to sell this surplus fat as fresh or acidified cream, both of which are more profitable than butter. In rural areas, however, the market for cream is often quite small and the cheese unit is obliged to make butter, a product of longer shelf life.

If the cream is to be sold as it is, it should be pasteurized before packaging. Experience has shown that a pasteurization temperature of approximately 95 - 98°C for 30 seconds destroys germs satisfactorily and inactivates enzymes while preserving the organoleptic qualities of the cream. After pasteurisation, the cream is packaged in plastic bags or tubs and kept in a refrigerator. The fresh cream is usually sold with a fat content of about 40 percent. Acidified cream is sold with a lower fat content - 30 - 35 percent.

To make butter, cream is cooled to the lowest possible temperature. The cream is then stored until enough is obtained to make into butter. Cream stored in this way acidifies automatically, after which it is churned.

After filling the churn, the following steps are taken:

— The churn is rotated at 25 - 35 revolutions per minute (rpm) for 5 minutes.

— The churn is stopped and if necessary the gases released.

— The butter is churned again at 25 - 35 rpm for 35 to 45 minutes.

— The buttermilk is poured off into plastic pails.

— The same amount of cold water as the amount of buttermilk which has been removed is added, and the mixture churned at 10 - 15 rpm for 5 minutes.

— Water is drawn off.

— Rotation at 10 - 15 rpm for 10 to 20 minutes.

— The butter is removed.

— The butter is worked with a butter worker for 5 minutes, compressed into a butter mould, and packaged in greaseproof paper. Alternatively, after working, it is placed in plastic tubs.

The butter can, if required, be salted as it is worked. This operation may be regarded as the standard butter-making process since a number of variations are possible. Non-acidified cream can be churned, in which case the product is called sweetcream butter. Some butter-makers wash the butter twice and in some cases the butter is not worked at all. The larger butter units usually pasteurize the cream, followed by the re-inoculation with selected lactic starter cultures for ripening. This method produces a cultured butter commonly called lactic butter.

Buttermilk

- Buttermilk is a by-product of the butter-making process. It is highly nourishing, hence the interest in promoting the product for human consumption.
- Buttermilk quality depends greatly on the butter-making technique used.
- Buttermilk can be packaged after only one filtration. It should be packaged in plastic bags.
- Buttermilk can also be ripened. While the buttermilk is still in the pail, cheese-making starter cultures (approximately 2 percent) are added, and the product left overnight at room temperature before packaging.

Yoghurt

Steps in yoghurt-making are as follows:

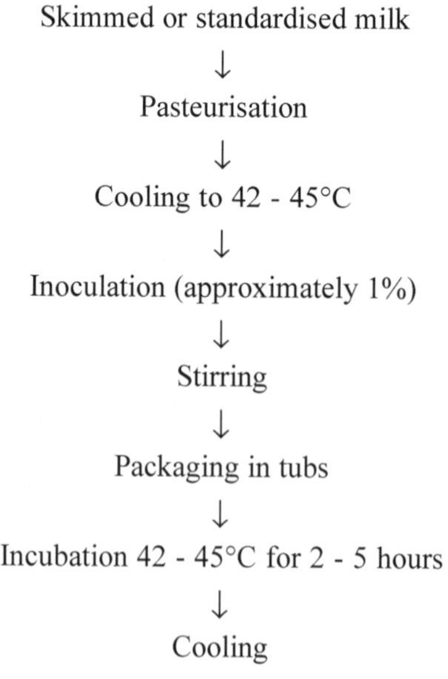

Pasteurisation and cooling are done in the cheese vat, or in a pan for smaller quantities. For yoghurt-making, the dairy unit needs to have a large refrigerator or small cold storage area—this also applies to cream and butter. Another essential investment item is an incubator, and perhaps a hand bottle-capper. Yoghurt is usually sold in plastic tubs or in cartons of 120 or 125 ml.

The yoghurt-making process increases the amount of cream available. Thus the decision to produce yoghurt should be preceded by a cost benefit analysis. Another technique is to let the yoghurt clot in the vat. It is then stirred prior to packaging.

Fermented Milk

The technology is very simple:

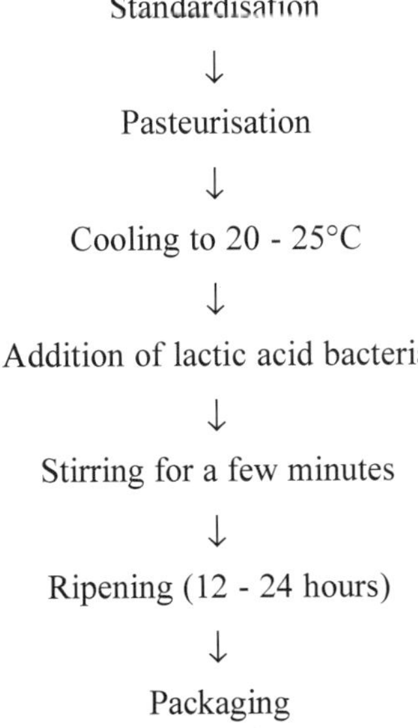

This technology corresponds to the traditional process, and the product is in demand as the base used in preparing "porridge". The processing unit's main consideration in this technique is that the milk is pasteurised and that both quality and hygienic standards are met.

Fermented milk being sold directly at the processing unit does not need to be packaged. The fermented milk is sold from the can, and is ladled into containers brought by local purchasers. No specific equipment is required to make fermented milk. It is, however, advisable to allow for a large refrigerator.

Dahi

This milk product is of major importance in the Indian sub-continent. It is a yoghurt-like product made in India and neighbouring countries. It is the most important fermented milk product used in India from times immemorial. The scale of production ranges from household level to industrial scale including preparation in halwai's milk shops in urban areas. Cow or buffalo milk or a mixture of the two is used. It is boiled and sometimes concentrated before addition of the starter which is usually a portion of the previous day's dahi or buttermilk.

Dahi has a milk pleasant flavour and a clean acid taste. It has a yellowish creamy-white colour when made from cow milk and a creamy-white colour when made from buffalo milk. It has a smooth and glossy surface. The body is firm but not hard and free from gas holes. Dahi is widely consumed all over India and the neighbouring countries including the Himalayan region, either plain, sugared or salted. The sweetened concentrated form of dahi consumed in Bengal is known as mishti doi i.e. sweet dahi.

Mishti Doi

A sweetened variety of dahi known as mishti doi, mishti dahi, lal dahi (red dahi) or payodhi in the eastern region of the Indian sub-continent is very popular. Cane sugar (6.0 – 6.5%) is added to the milk before boiling. Artificial colour, caramel and jaggery may also be added. The milk is cooled to 40–45°C and incubated for 12 – 15 hours.

Lassi

Dahi is converted into this refreshing beverage by stirring and adding a small quantity of water. It is best consumed chilled, and either sweetened or salted. It is a preferred drink in the northern parts of the sub-continent particularly the Punjab and Haryana. It is known to induce sleep particularly after consumption during the summer afternoons. Aseptically packed long life lassi has recently been introduced in India.

Shrikhand

Shrikhand or Sikarni, as it is known in Nepal, is made from concentrated dahi with a sweet and sour taste. It is a semi-soft whole milk product resembling sweetened quarg or quark produced in Germany. Shrikhand is traditionally made at home in western India. The name shrikhand is derived from the Sanskrit work shikharini. Dahi is placed in a muslin cloth and drained for 4–8 hours to reduce the whey content and produce a solid mass called chakka or maska. Chakka is mixed with the required amount of sugar, condiments and flavour to produce shrikhand. An industrial process for the manufacture of chakka and shrikhand has been developed by the National Dairy Development Board of India.

Shrikhand Wadi

This product is obtained by further concentration of shrikhand as prepared above by heating in an open pan over a direct fire until it forms a hard mass.

Shrikhand wadi has the following composition.

Moisture	5–6(per cent)
Fat	7–8
Protein	8–10
Lactose	15–17
Ash	0.75–0.80
Sugar	63–65
Lactic acid	1.0–1.2

Chhaas (Buttermilk)

Buttermilk produced by the churning of soured milk (dahi) is known as chhaas or chhach and as mahi in Nepal. The fat content is usually from 1–2 per cent and it is rich in protein and lactose. Chhaas is mostly consumed in the household and surplus is fed to cattle.

Kadhi.

This product is made from chhaas by a recipe which varies from region to region. A blend of spices, of which the common ingredients are salt, black pepper, green chillies, turmeric, coconut, and ground cumin are added with a small amount of Bengal gram flour to an appropriate quantity of chhaas and the mixture is brought to boiling point. It is then served hot with rice. In some regions of the country, small balls made out of besan dough and fried in oil are added to kadhi and served as a curry.

Cheese Products

Melting emulsification has traditionally facilitated the incorporation of sub-standard cheeses, as well as cheese trimmings, into products for sale as processed cheese. The usual presentation for processed cheese is as a spread, packaged in plastic rolls or aluminium-wrapped in bite-sized portions.

While the latter requires costly equipment to mould and package the cheese, processed cheese spread in plastic rolls offers an attractive approach for small dairies where second-quality cheeses, or cheeses with a manufacturing defect are readily available. The product can be packaged in glass or plastic tubs, or otherwise wrapped in aluminium.

Steps in processed-cheese making:

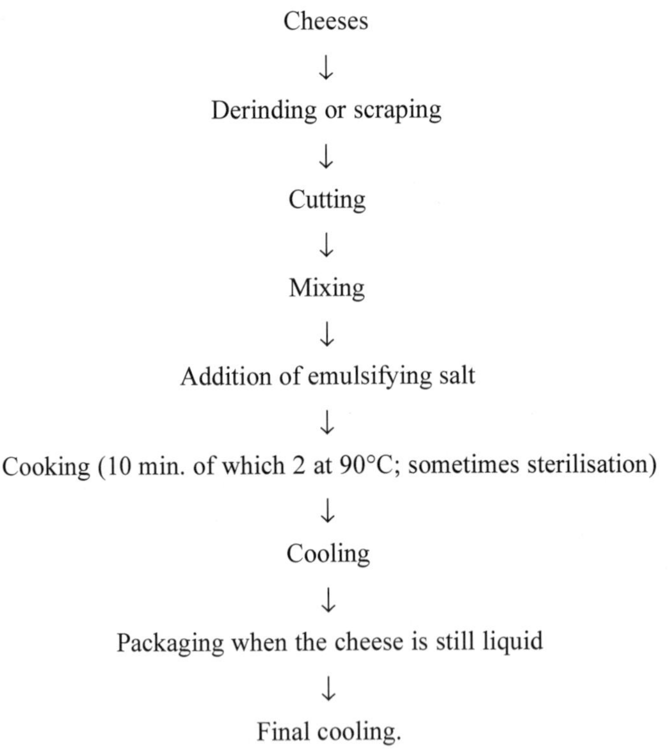

Commercial processed cheese preparations are highly diversified, as it is possible to add a variety of ingredients: flavourings, mushrooms, nuts, meats, etc. Packaging while the cheese is liquid, allows great variety of shapes and weights in the finished product. No specific equipment is required to make processed cheese, the only indispensable item being a pot in which to melt the cheese.

Paneer

Paneer consists mainly of acid-coagulated milk solids and is used extensively as an ingredient in many cooked vegetable preparations in Northern India, Pakistan, Afghanistan and Nepal. Paneer making is confined to the North-west frontier regions of the Indian sub-continent. It is produced at small scale and industrial level. Cow, buffalo or mixed milk may be used but buffalo milk is preferred.

The milk is boiled and the coagulation is simultaneously effected by adding the required amount of coagulant acid in a thin stream, within a minute, and mixing it into the milk with a stirrer. Draining is begun when the whey is clear. On a commercial scale, Paneer is processed mechanically into blocks in hoops by putting weights on the hoops (approx 2–3 kg per sq cm for 15–20 min). Drained and pressed curd is cut into suitable sizes and immersed in chilled water for 3–4 hours to make it firm. It is usually sold in pieces without packaging.

An industrial-scale process has been developed by the NDDB. Milk is heated to 85°C through a plate heat exchanger and pumped to a cheese vat and cooled to 75°C. Citric acid solution is added and mixed with the milk to form a coagulum. The curd is left to settle for 10–15 min without agitation. The whey is drained off. Curd is filled into cheese hoops lined with muslin cloth. Pressing of the curd for 10–15 min at a pressure of 3 kg per sq cm. Pressed curd blocks are place in pasteurized cold water at 4°C for 3 hours. The cooled blocks of paneer are cut into 200 g or 500 g portions which are wrapped in vegetable parchment paper before being placed in HDPE or LDPE bags and heat sealed ready for sale.

In India, paneer must meet the following legal requirements:

— The yield of paneer depends on the quality of milk. It is generally 18 to 20 per cent of the weight of the milk used for its preparation.

— 'White' paneer is a staple food of nomads in Afghanistan. It is traditionally consumed in the northern regions of the Indian sub-continent with dry fruits and nuts as a dessert.

— Paneer is also the Hindu name of the seeds of Withania coagulans, the basis of a vegetable coagulant that yields a bitter curd.

— Curdled milk products obtained by the admixture with sour milk, pieces of a creeper called Putika, the bark of Palasa trees or Kuyala (Jukuke) was known to the ancient Indians.

However the curdled milk product, paneer, seems to have been introduced into India from the Middle East perhaps by Persian and Afghan invaders. A unique Iranian nomadic cheese is called paneer Khiki. This cheese was originally developed by the well-known Bakhtiari tribe which resided in Isfahan (in summer) and Shiraz (in winter). The word khiki means skim.

Rennet from the goat or sheep was used to make the paneer, hence the name. When salted it known as paneer-e-shour. It is only in the past four decades that consumption of paneer has spread to other parts of India. It enjoys the status of haute cuisine amongst Indian vegetarian cooking.

Surti Paneer

The name of this cheese is derived from the town of Surat in western India where it was probably first prepared and marketed. Once a popular product, very little of it is marketed today. It is a soft cheese prepared from buffalo milk with crude rennet, salted and kept steeped in acid whey for 2–3 days. Surti paneer should have a fairly firm body and smooth texture with no internal cracks. It has a slightly salted, milk acid-curd flavour.

Bandel Cheese

Bandel cheese is an indigenous unripened, salted soft variety of cheese made in perforated pots. It is similar to surti paneer but made from cow's milk. It is available in and around Bandel, a Portugese colony in eastern India, and seems to have derived its name from it. The cheese is formed into a flattened circular shape and is ready for immediate sale.

Dacca

This cheese is available in the eastern region. It is similar to bandel but differs from it in that the finished flat round cheeses are smoked in a fire.

Chhanna

It consists of acid coagulated milk solids used for the preparation of many milk based sweets. It differs from Paneer in that no pressure is applied to remove the whey. Chhanna is widely used in the eastern parts of India and Bangladesh. Cow milk is preferred since it yields a soft bodied and smooth textured product. Both these characteristics are suitable for the production of high grade chhanna sweets.

Buffalo milk produces a chhanna with a slightly hard body, a greasy and coarse texture, and does not produce good quality chhanna sweets. Chhanna has the same legal requirements as paneer in India, i.e. a maximum moisture content of 70 per cent and a minimum content of milk fat in dry matter of 50 per cent.

Chhanna from cow milk is light yellow in colour, has a moist surface, soft body and smooth texture. Chhanna derived from buffalo milk is whitish in colour. Both have a pleasant sweetish, mildly acid taste. Buffalo milk yields a larger amount of chhanna. About 100,000 metric tons are produced annually in India. Chhanna is also produced in rural milk sheds and transported by road and rail to larger urban conglomerates in wicker baskets which allow further drainage of whey. Chhanna produced in this way is used for the preparation of Sandesh.

Chhanna-based Sweets: Rasogolla

This sweet is of recent origin having been developed in 1868 by an enterprising Calcutta sweetmeat maker Nobin Chandra Das. It is prepared using fresh and soft-chhanna. In the form of balls 30 mm in diameter with a typical spongy body and smooth texture. Stored and served in sugar syrup. Freshly-made chhanna is squeezed by hand in a muslin cloth to remove as much whey as possible. 1–4 per cent of the wheat flour/semolina is mixed with the chhanna in a container and kneaded thoroughly by hand to make a dough.

The dough is portioned and rolled into balls of about 15 mm diameter having a smooth surface with no cracks—1 kg of chhanna yields 90–100 rasogollas. The dough balls are cooked in a specially prepared whey based medium for about 15 minutes. For chhanna made from cow milk, cooking medium with sugar is preferred, and for all other types of chhanna, cooking medium without sugar is preferred.

After the cooking is complete, the balls are transferred to a container with water at 30–35°C for texture stabilisation and colour improvement of the balls. After 5–10 min of texture stabilisation in water, the texture stabilised balls are transferred to sugar syrup. The desired sugar syrup concentration in the final product is 45–50 per cent. This is achieved by dipping the texture-stabilised balls first in 35–40 per cent sugar syrup for 1–2 hours, followed by a second dipping in 58–60 per cent sugar syrup. The product finally acquires the desired sugar concentration after equilibration between the sugar syrup inside and outside the balls is achieved.

Lalmohan

A product similar to gulabjamun but is made from chhanna and is lighter in colour. Chhanna is mixed with 2–3 per cent wheat flour and kneaded into a uniform dough. The dough is rolled into small balls and deep fried in ghee until light brown in colour. The balls are transferred to a 60 per cent sugar syrup and allowed to soak for a few hours before being served.

Other Milk-Based Products

Khoa

This product is obtained from cow, buffalo or mixed milk by thermal evaporation of milk to 65–70 per cent solids in an open pan. A five times concentration of milk is normally required for the production of khoa. Khoa, also khawa or mawa, is used as a base material for a variety of Indian sweets. Its origin is not known but it has been prepared for centuries in India as the base material for sweets. About 600,000 metric tons of khoa is produced annually in India alone. It is made by the traditional method by milk traders and halwais.

Khoa preparation has been the easiest way of preserving rurally-produced milk in the flush season. In many places khoa manufactured in January—February is cold-stored for use in the summer season. Such khoa acquires a green colour due to mould growth on the surface. It is therefore known as hariyali (green khoa). This khoa is preferred for the preparation of gulabjamun as it gives a grainier texture to the product. This khoa, on removal from the cold store is immediately mixed with flour and made into gulabjamuns. Hariyali khoa, if left at room temperature for long, starts to smell and breaks down physically. Because of this it is converted into products immediately.

Legal requirements state that khoa contains a minimum of 20 per cent milk fat. The Bureau of Indian Standards has laid down the following specifications for khoa. Milk of high acidity produces a granular khoa known as danedar. Khoa has a uniform whitish colour with just a tinge of brown, a slightly oily or granular texture, and a rich nutty flavour which is associated with a mildly cooked and sweet taste due to the high concentration of lactose.

Buffalo milk is preferred for khoa making because it yields a whiter product with a soft, loose body and a smooth granular texture which makes it suitable for the preparation of high-grade khoa sweets. A minimum of 4 per cent fat for cow milk and 5 per cent fat for buffalo milk is necessary to obtain a desirable body and texture in khoa. Lower levels of fat result in undesirable hard body and coarse texture. The traditional trade usually pays for milk on the basis of the yield of khoa. Cow milk usually yields 18 per cent of khoa. The yield from buffalo milk is usually 20 per cent.

Peda

The quantity of peda produced in India exceeds any other indigenous milk-based sweet using khoa as the raw material. Peda or doodh peda is prepared on a small scale by halwais using khoa as the base material mixed with sugar and flavourings. A similar product which is very popular in Nepal is called gundpak.

Peda is usually packed in paperboard cartons with a parchment paper of grease proof paper liner. it is usually sold through confectionery shops Peda is whitish yellow in colour and has a coarse grainy texture. Kesar (saffron) peda is one of the preferred pedas in which saffron is used for added flavour and colour.

Other Khoa-based Sweets

The methods of preparation and various features of other khoa-based sweets—burfi, kalakand, gulabjamun, and kalajanum or kalajam.

Condensed Milk-based Products

This sub-group of milk-based products includes rabri, khurchan, basundi, kheer and palpayasam. Features of preparation are given below.

Rabri, Tar (in Nepal): A specially prepared concentrated and sweetened whole milk product containing several layers of clotted cream. It is a sweet by itself and is not much used as a component of other sweets. It is produced in Northern and eastern regions of India normally from buffalo milk is normally used since it produces a more creamy and chewy consistency. In comparison to cow milk the higher fat and casein contents of buffalo milk contribute to the formation of a greater volume of creamy layer early in the evaporation process.

Milk (3–4 kg) is heated in a fairly shallow pan over an open fire and allowed to simmer, 5–6 per cent of sugar is added and evaporated to one eighth of the original volume. The preparation time is about 25–40°C minutes depending on the rate of boiling. The finished product consists of non homogeneous flakes partly covered by and partly floating in sweetened condensed milk. By heating the concentrate slightly at the end, a more homogeneous chewy-textured mass is obtained. The following composition relates to Rabbri prepared in Nepal.

Khurchan: A concentrated, sweetened whole milk product, similar to Rabri. It is used for direct consumption. It is produced in the Northern region of India almost exclusively from bufflo milk as it gives a higher yield than cow milk. The final produce has a slightly cooked flavour, which is relished.

Basundi: A concentrated milk to which sugar, flavours and nuts are added. The product is served chilled as a dessert. The origin of the product is not known but it has been traditionally prepared for centuries in the western part of India as a dessert, served on special occasions such as weddings. About 25,000 metric tons of basundi are produced annually in India from cow and buffalo milk on a small-scale.

Milk, in a shallow pan is boiled on a low flame. The heat coagulated film that appears on the surface of the milk is collected and spread on the sides of the vessel. The volume of milk is reduced to 50 per cent of it's original volume. The pan is removed from the fire and sugar is added along with nuts and flavours. The mass is mixed until the sugar is dissolved. The product is cooled and served chilled. Basundi looks like condensed milk with flakes. It has a light brown colour with thin flakes in a thick fluid. It has a pleasant flavour similar to condensed milk. The cooked flavour is relished by the consumer.

Kheer: A sweetened product of thick consistency resembling rice pudding commonly consumed in the West. The product is prepared for immediate consumption. It is produced in northern, western and central regions of the Indian sub-continent. This product is widely consumed in the regions mentioned above. In Nepal, the housewife perpares kheer by concentrating whole milk in open pans with the addition of sugar (6–8 per cent), rice (6–7 per cent), ghyu, cashew nuts, cardamon and other spices.

Palpayasam: A sweetened product; similar to kheer and resembling a rice pudding produced in Southern India. Vermicelli or semolina may be substituted for rice, fruits like jack fruit are optional.

Frozen Milk Products: Kulfi or Malai Kulfi.

This indigenous ice cream is based on milk and is popular in the hot summer. It is frozen in small containers. The preparation of kulfi involves concentration of a milk and sugar mixture to 50 per cent volume. It is cooled before addition of cooled cream, crushed nutes and selected flavourings. The milk is added to moulds and frozen in a vessel containing an ice and salt mixture with a 1:1 ratio. The Bureau of Indian Standards has laid down the following specifications for kulfi.

Importance of Milk Products

With an annual milk production of 46 million metric tons in 1986–87, India ranks third in the world after the Soviet Union and the United States of America. Milk production is expected to further surge forward in the coming years to cross the 61 million metric tons mark by 1995. The target for 2000 AD is 70 million metric tons. Milk and milk products play a vital role in the agricultural economy, being the second largest agricultural product in India.

In 1984–85 the value of milk and its products exceeded Rs.100,000 million, ranking after rice, but before wheat. Equally important, if not more so, is the role of dairying in providing sustenance to millions of farmers, constituting 75 per cent of the total population in some 80 million farm households distributed over 550,000 villages, constituting the bulk of rural poor, with an annual income of less than Rs. 3,800 per family. Milk provides both nutrition and supplementary income to these weaker sections.

Over 5 million farm families, in 49,000 village milk producer's cooperatives, sell on an average some 8 million litres of milk every day, after retaining some 30 per cent of it for their own consumption. Sale of milk fetches them about Rs. 30 million per day aggregating Rs. 10,000 million per year. India alone produces some 650,000 metric tons of ghee valued at Rs. 32,000 million. The value of resultant lassi is Rs. 10,000 million, some 600,000 tonnes of khoa valued at Rs. 18,000 million is produced in India along with some 100,000 metric tons of chhanna valued at Rs. 3,000 million.

The value of khoa and chhanna produced in India is probably twice the value of all the milk handled by the organised sector in the country. The traditional dairy products sector in India, like its agricultural economy, is grossly under managed. However, it provides economic opportunities that even the western dairy world would be envious of. The value of khoa and chhanna-based sweets could possibly exceed US $ 4 billion. In the absence of reliable data, the above figures are only rough estimates, but highlight the significance of the traditional dairy products in the national economy.

The composition of milk makes it an ideal balanced food for humans especially infants and its importance as a supplement to the average diet cannot be over emphasised. Traditionally in areas where milk production is abundant, milk and milk products are regularly consumed by almost all sections of the population. For example, the average Punjabi diet can compare well with some of the best diets in the world. However, the same cannot be said of the major sections of the population.

Recognising the proper role that milk can play in the nutrition of the people, efforts are being made to increase milk production significantly. Supplementary feeding programmes for infants and expectant mothers, and school children, have always included milk powder as one of the ingredients. There is hardly any major difference in the nutritive values of cow and buffalo milk except for the greater calorific value of buffalo milk due to its higher fat content.

Although 46 per cent of the milk produced in the country is consumed as liquid milk and as such milk plays an important role in the national diet, there is considerable need and scope for increased consumption of milk. The expenditure elasticity of demand for milk is very high in India, 1.46 for the rural population and 1.3 for the urban population.

The daily allowances of nutrients for an Indian adult male recommended by the Indian Council of Medical Research are given below. The average diet of the poorer sections of the population is deficient in several nutrients and most of these can be made up by supplementing the diet with milk. As against the recommended level of 200 ml of milk, the average per capita intake was 168 ml, in 1988. Except in the case of high and middle income groups it is less than the recommended levels.

Milk plays a major role as a source of proteins in the average Indian diet contributing some 10 per cent of the protein intake. These data are indicative of the important part that milk plays in the nutrition of the population. In India most milk is boiled before consumption. It is to this practice that the absence of milk-borne diseases in India is to be mainly attributed. Heating to first boil results in destroying most of the organisms. Denaturation of proteins as well as its flocculation due to the neutralisation of the electric charges occurs to some extent on boiling milk.

A partial precipitation of calcium salts and phosphates also occurs, the diffusible calcium being reduced from 26 per cent to 20 per cent. Among the vitamins in milk, A is the most resistant, and C the most vulnerable to heat treatment. While hardly any vitamin A is destroyed by boiling, about 22 per cent of vitamin C is lost when milk is boiled. The loss of vitamin C is dependant both on the time of treatment and the exposure to light. A slight reduction in the thiamine (B1) content of milk occurs. Riboflavin (B2) is hardly affected. The availability of calcium and vitamins (except vitamin C) is not affected by boiling.

Most of the enzymes of milk are destroyed during boiling and the digestibility of milk increases. Hot milk is widely consumed before going to bed as a nightcap. The milk is usually flavoured with condiments such as almonds, cardamom, dry dates etc. In the Indian households the life of milk is extended from 12 to 24 hrs by repeated boiling. The simplest way of preserving milk for human consumption in a tropical country is to allow it to sour with the aid of lactic cultures, checking putrefactive changes while giving to milk an acid taste which is particularly refreshing in a hot climate.

The product thus achieved, dahi, is widely consumed in the country along with meals. The digestibility of milk constituents improves. Dahi can also be consumed by people who suffer from lactose intolerance. Almost every household in the country consumes dahi. Due to fermentation of milk a greater amount of phosphorus and calcium is made available to the digestive system by their precipitation in the lower intestines due to the acid condition induced by Lactobacillus sp.; and the consumption of sour milk also results in increased efficiency of the body to cope with a sudden influx of lactic acid in the system.

It is reported that when the food is supplemented with 250 g of dahi a day, the status of thiamine improves. Dahi also increases the pyruvic acid and the lactic acid among children on a typical poor rice diet. Thus, dahi in its different forms, lassi, kadhi, shrikhand etc. also contributes significantly to the average diet. Makkhan and ghee contribute as much as one third of the fat in the Indian diet.

Ghee is produced mainly for consumption directly as food and as an ingredient of food preparations including sweets. Over the centuries Indians have cultivated a liking for the aroma and flavour of ghee, and a preference for its use over vegetable oils, the other traditional cooking medium for the preparation of specific food items.

The vegetarian habits of many Indians preclude from their diet hard animal fats such as tallow or lard used in the West and thus ghee forms an important source of fat in an otherwise vegetarian diet. For most uses, its wholesome flavour is the chief attraction. For table use it is served in melted form and mixed with rice or lightly smeared on chapatis. It is widely used for shallow frying and deep frying of food materials. Innumerable Indian sweetmeats based on cereals, milk solids, fruits and vegetables are cooked, by preference, in ghee.

Buttermilk or lassi as described earlier is a by-product in the preparation of makkhan. It is estimated that about 55 kg of buttermilk is produced for every kg of ghee. While most of this is consumed by the villagers and their families, a good quantity is either given away or fed to cattle. The reason for this is the lack of market value for the product in rural parts. Buttermilk is rich in milk protein and calcium and forms a valuable human food.

Ghee and makkhan are important carriers of vitamins A, D, E & K. They also contain small amounts of essential fatty acids e.g. arachidonic and linoleic. Considerable losses of Vitamin A and carotene occur during cooking, the loss of the latter being more rapid. Below 125°C Vitamin A is fairly stable but above this temperature it is rapidly destroyed. It is found that 10–20 per cent of carotene is lost during the normal cooking operations.

Marketing of the Dairy Products

Marketing of the indigenous dairy products is as traditional as are the products. Halwais, the traditional sweetmeat sellers, produce and sell these products in all urban and semi-urban areas of the region. Halwais have prospered over the years as the products have high margins of profits.

The festival season (October—November) sales in many areas account for 30–40 per cent of the annual sales of traditional milk-based sweets. Most sales are made across the counter for ready cash. Kulfi (ice cream) is usually sold from door to door by hawkers and some manufacturers also offer kulfi as a part of their wide range of ice creams.

Ghee is usually marketed in the traditional markets (mandies) by those who collect it from the villages and refine it further to remove all moisture. From the mandies, it goes to retail outlets. Ghee is usually branded by large traders. Large ghee mandies exist in Hathras, Khurja, Porbandar, Guntur and Erode. Jodhpur is probably the largest ghee trading centre in India. Some manufacturers now advertise ghee on the national television network. Organised dairies brands of ghee now fetch a good price as the quality is considered to be guaranteed.

Expanding markets: Some of the traditional halwais have more than one outlet in a city. Some have started canning rasogollas and gulabjamuns. Bikaner in the west has emerged as a large custom processing and manufacturing centre for rasogollas. Halwais in Bikaner execute large orders and provide custom labelling for canned rasogollas which are manufactured in the traditional way but do not reach the quality of the traditional product.

Modernisation of marketing efforts: The most modern plant manufacturing traditional dairy products is the Sugam cooperative dairy at Baroda marketing its products through a large network of 150 retail outlets in the city. The Sugam Dairy uses the traditional grocery/ general stores that have a refrigerator to market its products. The product range includes shrikhand, gulabjamuns, peda and lassi apart from flavoured milks. The dairy has the highest turnover of a single unit marketing traditional dairy products. Other dairies are active in marketing traditional milk products.

Cleaning and Disinfection

Cleaning consists of removing all visible or invisible dirt from the surface. This surface can thus be described as clean. Disinfection involves eliminating or killing micro-organisms. Effective cleaning is absolutely essential for the equipment, installations and premises used for cheese-making, and the cheese-maker needs to pay special attention to this item.

After each use, all equipment and utensils: pails, cans, filters, pans, trays, tables, ladles etc., must be vigorously and meticulously cleaned. This is essential for successful cheese-making as the equipment is the main source of contamination by harmful germs. Simply rinsing with cold or lukewarm water is not sufficient. This is because a very thin film of residues or wastes from the clotting process sticks to the surfaces of the equipment which has come into contact with milk, whey or curd.

To eliminate these residues, an alkaline or acid detergent solution must be added accompanied either by vigorous brushing or agitation of the soaking solution to remove the residual film or wastes. It is recommended that the equipment be soaked immediately after use in a vat filled with water.

The layout and design of the dairy unit should be planned in such a way as to allow those responsible for processing operations to work under the confortable, safe and hygienic conditions.

References

Atkins, P.J., 2000. "The pasteurization of England: the science, culture and health implications of milk processing, 1900-1950", In: Smith, D. & Phillips, *Journal of Food, science and regulation in the 20th century*, Routledge,

Gegner, L., 2001."Value-Added Dairy Options", *ATTRA* (Appropriate Technology Transfer for Rural Areas), Fayetteville, Ark., Aug.

Robinson, R. K., 1986. *Modern Dairy Technology*, Volume 1. Elsevier AppliedScience Publishers, London.

Varnam, A. H. and Sutherland, J. P., 1997. *Milk and Milk Products:Technology, Chemistry and Microbiology*, Chapman and Hall, London, World Bank.

6

Meat Processing Technology

Meat consumption in developing countries has been continuously increasing from a modest average annual per capita consumption of 10 kg in the 1960s to 26 kg in 2000 and will reach 37 kg around the year 2030 according to FAO projections. This forecast suggests that in a few decades, developing countries' consumption of meat will move towards that of developed countries where meat consumption remains stagnant at a high level.

The rising demand for meat in developing countries is mainly a consequence of the fast progression of urbanization and the tendency among city dwellers to spend more on food than the lower income earning rural population. Given this fact, it is interesting that urban diets are, on average, still lower in calories than diets in rural areas. This can be explained by the eating habits urban consumers adopt. If it is affordable to them, urban dwellers will spend more on the higher cost but lower calorie protein foods of animal origin, such as meat, milk, eggs and fish rather than on staple foods of plant origin. In general, however, as soon as consumers' incomes allow, there is a general trend towards incorporating more animal protein, in particular meat, in the daily diet. Man's propensity for meat consumption has biological roots. In ancient times meat was clearly preferred, consequently time and physical efforts were invested to obtain it, basically through hunting. This attitude contributed decisively to physical and mental development of humankind. Despite the growing preference in some circles for meatless diets, the majority of us will continue eating meat. It is generally accepted that balanced diets of meat and plant food are most effective for human nutrition.

Quantitatively and qualitatively, meat and other animal foods are better sources of protein than plant foods (except soy bean products). In meat, the essential amino acids – the organic acids that are integral components of proteins and which cannot be synthesized in the human organism – are made available in well balanced proportions and concentrations. As well, plant food has no Vitamin $B_{12;}$ thus animal food is indispensable for children to establish B_{12} deposits. Animal food, in particular meat, is rich in iron, which is of utmost importance to prevent anemia, especially in children and pregnant women.

In terms of global meat production, over the next decade there will be an increase from the current annual production of 267 million tons in 2006 to nearly 320 million tons by 2016. Almost exclusively, developing countries will account for the increase in production of over 50 million tons. This enormous target will be equivalent to the levels of overall meat production in the developing world in the mid-1980s and place an immense challenge on the livestock production systems in developing countries.

The greater demand for meat output will be met by a further shift away from pastoral systems to intensive livestock production systems. As these systems cannot be expanded indefinitely due to limited feed availability and for environmental reasons, other measures must be taken to meet growing meat demand. The only possible alternatives are making better use of the meat resources available and reducing waste of edible livestock parts to a minimum.

This is where meat processing plays a prominent role. It fully utilizes meat resources, including nearly all edible livestock parts for human food consumption. Meat processing, also known as further processing of meat, is the manufacture of meat products from muscle meat, animal fat and certain non-meat additives. Additives are used to enhance product flavour and appearance. They can also be used to increase product volume. For specific meat preparations, animal by-products such as internal organs, skin or blood, are also well suited for meat processing. Meat processing can create different types of product composition that maximizes the use of edible livestock parts and are tasty, attractive and nourishing.

The advantage of meat processing is the integration of certain animal tissues (muscle trimmings, bone scraps, skin parts or certain internal organs which are usually not sold in fresh meat marketing) into the food chain as valuable protein-rich ingredients. Animal blood, for instance, is unfortunately often wasted in developing countries largely due to the absence of hygienic collection and processing methods and also because of socio-cultural restrictions that do not allow consumption of products made of blood. While half of the blood volume of a slaughtered animal remains in the carcass tissues and is eaten with the meat and internal organs, the other half recovered from bleeding represents 5-8 percent of the protein yield of a slaughter animal. In the future, we cannot afford to waste such large amounts of animal protein. Meat processing offers a suitable way to integrate whole blood or separated blood fractions (known as blood plasma) into human diets.

Meat processing technologies were developed particularly in Europe and Asia. The European technologies obviously were more successful, as they were disseminated and adopted to a considerable extent in other regions of the world – by way of their main creations of burger patties, frankfurter-type sausages and cooked ham. The traditional Asian products, many of them of the fermented type, are still popular in their countries of origin. But Western-style products have gained the upper hand and achieved a higher market share than those traditional products.

In Asia and Africa, there are a number of countries where meat is very popular but the majority of consumers reject processed meat products. This is not because they dislike them but because of socio-cultural reasons that prohibit the consumption of certain livestock species, either pork or beef depending on the region. Because processed products are mostly composed of finely comminuted meat, which makes identifying the animal species rather difficult, or are frequently produced from mixes of meat from different animals, consumers stay away from those products to avoiding eating the wrong thing. But when the demand for meat increases and a regular and cost-effective supply can only be achieved by fully using all edible livestock parts, consumers will need to adjust to processed meat products, at least to those where the animal source can be identified. Younger people already like to eat fast-food products such as beef burgers or beef frankfurters. Outlet chains for such products and other processed meat products will follow when the demand increases.

Sources of Meat, Fat and Animal By-products

Meat, fat and other carcass parts used as raw materials for the manufacture of processed meat products are mainly derived from the domesticated animal species cattle, pigs and poultry and to a lesser extend from buffaloes, sheep and goats. In some regions other animal species such as camels, yaks, horses and game animals are used as meat animals but play only a minor role in meat processing.

In this context, meat can be defined as "the muscle tissue of slaughter animals". The other important tissue used for further processing is fat. Other edible parts of the slaughtered animal and often used in further processing are the internal organs (tongue, heart, liver, kidneys, lungs, diaphragm, oesophagus, intestines) and other slaughter by-products (blood, soft tissues from feet, head).

A special group of internal organs are the intestines. Apart from being used as food in many regions in particular in the developing world, they can be processed in a specific way to make them suitable as sausage casings. Some of them are eaten with the sausage; others are only used as container for the sausage mix and peeled off before consumption.

The skin of some animal species is also used for processed meat products. This is the case with pork skin and poultry skin, in some cases also with calf skin.

Characteristics of Meat

In general, meat is composed of water, fat, protein, minerals and a small proportion of carbohydrate. The most valuable component from the nutritional and processing point of view is protein.

Protein contents and values define the quality of the raw meat material and its suitability for further processing. Protein content is also the criterion for the quality and value of the finished processed meat products.

Water is a variable of these components, and is closely and inversely related to the fat content. The fat content is higher in entire carcasses than in lean carcass cuts. The fat content is also high in processed meat products where high amounts of fatty tissue are used.

The value of animal foods is essentially associated with their content of proteins. Protein is made up of about 20 aminoacids. Approximately 65% of the proteins in the animal body are skeleton muscle protein, about 30% connective tissue proteins (collagen, elastin) and the remaining 5% blood proteins and keratin (hairs, nails).

Changes of pH

Immediately post-mortem the muscle contains a small amount of muscle specific carbohydrate, called glycogen (about 1%), most of which is broken down to lactic acid in the muscle meat in the first hours (up to 12 hours) after slaughtering. This biochemical process serves an important function in establishing acidity (low pH) in the meat.

Meat Colouring

The red pigment that provides the characteristic colour of meat is called myoglobin. Similar to the blood pigment haemoglobin it transports oxygen in the tissues of the live animal. Specifically, the myoglobin is the oxygen reserve for the muscle cells or muscle fibres. Oxygen is needed for the biochemical process that causes muscle contraction in the live animal. The greater the myoglobin concentration, the more intense the colour of the muscle. This difference in myoglobin concentration is the reason why there is often one muscle group lighter or darker than another in the same carcass.

Myoglobin concentration in muscles also differs among animal species. Beef has considerably more myoglobin than pork, veal or lamb, thus giving beef a more intense colour. The maturity of the animal also influences pigment intensity, with older animals having darker pigmentation. The different myoglobin levels determine the curing capability of meat. As the red curing colour of meat results from a chemical reaction of myoglobin with the curing substance nitrite, the curing colour will be more intense where more muscle myoglobin is available.

Water-holding Capacity

The water holding capacity (WHC) of meat is one of the most important factors of meat quality both from the consumer and processor point of view. Muscle proteins are capable of holding many water molecules to their surface. As the muscle tissue develops acidity (decrease of pH) the water holding capacity decreases.

Water bound to the muscle protein affects the eating and processing quality of the meat. To obtain good yields during further processing including cooking, the water holding capacity needs to be at a high level.

Water holding capacity varies greatly among the muscles of the body and among animal species. It was found that beef has the greatest capacity to retain water, followed by pork, with poultry having the least.

Tenderness and Flavour

Meat tenderness plays an important role, where entire pieces of meat are cooked, fried or barbecued. In these cases some types of meat, in particular beef, have to undergo a certain ripening or ageing period before cooking and consumption in order to achieve the necessary tenderness. In the fabrication of many processed meat products the toughness or tenderness of the meat used is of minor importance. Many meat products are composed of comminuted meat, a process where even previously tough meat is made palatable. Further processing of larger pieces of meat (e.g. raw or cooked hams) also results in good chewing quality as these products are cured and fermented or cured and cooked, which makes them tender.

The taste of meat is different for different animal species. However, it may sometimes be difficult to distinguish the species in certain food preparations. For instance, in some dishes pork and veal may taste similar and have the same chewing properties. Mutton and sometimes lamb has a characteristic taste and smell, which originates from the fat. Even small quantities of fat, e.g. inter- and intramuscular fat, may imprint this typical smell and taste on the meat, particularly of meat from old animals. Feed may also influence the taste of meat (e.g. fish meal). In addition, the sex of the animal may also give a special taste and smell to the meat. The most striking example is the pronounced urine-like smell when cooking old boar's meat. Meat fit for human consumption but with slightly untypical smell and flavour, which may not be suitable for meat dishes, can still be used for certain processed meat products. However, it should preferably be blended with "normal" meat to minimize the off-odour. Also intensive seasoning helps in this respect.

Animal Fats

Fatty tissues are a natural occurring part of the meat carcass. In the live organism, fatty tissues function as

— Energy deposits (store energy)

— Insulation against body temperature losses

— Protective padding in the skin and around organs, especially kidney and heart.

Fatty tissue is composed of cells, which like other tissue cells, have cell membranes, nucleus and cell matrix, the latter significantly reduced to provide space for storing fat. Fats, in the form of triglycerides, accumulate in the fat cells. Well fed animals accumulate large amounts of fat in the tissues. In periods of starvation or exhaustion, fat is gradually reduced from the fat cells.

For *processed meat* products, fats are added to make products *softer* and also for *taste* and *flavour improvement*. In order to make best use of animal fats, basic knowledge on their selection and proper utilization is essential.

Fatty tissues from certain animal species are better suited for meat product manufacture, fats from other species less or not suited at all. This is mainly for sensory reasons as taste and flavour of fat varies between animal species. Strong differences are also pronounced in older animals, with the well known example of fat from old sheep, which most consumers refuse. However, this aspect is to some extent subjective as consumers prefer the type of animal fat they are used to.

Nutritional Value of Meat and Meat Products

Proteins

The nutritional value of meat is essentially related to the content of high quality protein. High quality proteins are characterized by the content of essential aminoacids which cannot be synthesized by our body but must be supplied through our food. In this respect the food prepared from meat has an advantage over those of plant origin. There are vegetable proteins having a fairly high biological value for instance soy protein, the biological value of which is about 65% of that of meat. Soy protein concentrates are also very useful ingredients in many processed meat products, where they not only enhance the nutritional value but primarily the water binding and fat emulsifying capacity.

Fats

Animal fats are principally triglycerides. The major contribution of fat to the diet is energy or calories. The fat content in the animal carcass varies from 8 to about 20%. The fatty acid composition of the fatty tissues is very different in different locations. External fat ("body fat") is much softer than the internal fat surrounding organs due to a higher content of unsaturated fat in the external parts.

Vitamins

Meat and meat products are excellent sources of the B-complex vitamins. Lean pork is the best food source of Thiamine (vitamin B_1) with more than 1 mg / 100 g as compared to lean beef, which contains only about 1/10 of this amount. The daily requirement for humans of this rarely occurring vitamin is 1-1.5 mg. Plant food has no vitamin B_{12}, hence meat is a good source of this vitamin for children, as in their organisms deposits of B_{12} have to be established. On the other hand, meat is poor in the fat soluble vitamins A, D, E, K and vitamin C. However, internal organs, especially liver and kidney generally contain an appreciable percentage of vitamin A, C, D, E and K. Most of the vitamins in meat are relatively stable during cooking or processing, although substantial amounts may be leached out in the drippings or broth. The drip exuding from the cut surface of frozen meat upon thawing also contains an appreciable portion of B-vitamins. This indicates the importance of conserving these fractions by making use of them in some way, for example through direct processing of the frozen meat without previous thawing (which is possible in modern meat processing equipment). Thiamine (vitamin B_1) and to a lesser extent vitamin B_6 are heat-labile. These vitamins are partially destroyed during cooking and canning.

Minerals

The mineral contents of meat include calcium, phosphorus, sodium, potassium, chlorine, magnesium with the level of each of these minerals above 0.1%, and trace elements such as iron, copper, zinc and many others. Blood, liver, kidney, other red organs and to a lesser extent lean meat, in particular beef are good sources of iron. Iron intake is important to combat anaemia, which particularly in developing countries is still widespread amongst children and pregnant women. Iron in meat has a higher bio-availability, better resorption and metabolism than iron in plant products.

Meat Processing Techniques

Meat processing technology comprises the steps and procedures in the manufacture of processed meat products. Processed meat products, which include various different types and local/regional variations, are food of animal origin, which contribute valuable animal proteins to human diets. Animal tissues, in the first place *muscle meat* and *fat*, are the main ingredients, besides occasionally used other tissues such as *internal organs*, *skins* and *blood* or *ingredients of plant origin*.

All processed meat products have been in one way or another physically and/or chemically treated. These treatments go beyond the simple cutting of meat into meat cuts or meat pieces with subsequent cooking for meat dishes in order to make the meat palatable. Meat processing involves a wide range of physical and chemical treatment methods, normally combining a variety of methods. Meat processing technologies include:

— Cutting/chopping/comminuting (size reduction)
— Mixing/tumbling
— Salting/curing
— Utilization of spices/non-meat additives
— Stuffing/filling into casings or other containers
— Fermentation and drying
— Heat treatment
— Smoking

Equipment Used in Meat Processing

In modern meat processing, most of the processing steps can be mechanized. In fact, modern meat processing would not be possible without the utilization of specialized equipment. Such equipment is available for small-scale, medium-sized or large-scale operations. The major items of meat processing equipment needed to fabricate the most commonly known meat products are listed and briefly described hereunder.

Meat grinder (Mincer)

A meat grinder is a machine used to force meat or meat trimmings by means of a feeding worm (auger) under pressure through a horizontally mounted cylinder (barrel). At the end of the barrel there is a cutting system consisting of star-shaped knives rotating with the feeding worm and stationary perforated discs (grinding plates). The perforations of the grinding plates normally range from 1 to 13mm. The meat is compressed by the rotating feeding auger, pushed through the cutting system and extrudes through the holes in the grinding plates after being cut by the revolving star knives. Simple equipment has only one star knife and grinder plate, but normally a series of plates and rotary knives is used. The degree of mincing is determined by the size of the holes in the last grinding plate. If frozen meat and meat rich in connective tissue is to be minced to small particles, it should be minced first through a coarse disc followed by a second operation to the desired size. Two different types of cutting systems are available, the "Enterprise System" and the "Unger System":

The smallest type of meat grinder is the manual grinder designed as a simple stuffing grinder, i.e. meat material is manually stuffed into the feeder. For all these small machines the Enterprise cutting system is used with one star knife and one grinder plate. These machines are very common everywhere in food processing but their throughput and production capacity is limited due to the small size and manual operation.

The intermediate size meat grinder, also designed as a stuffing grinder, has orifice diameters up to 98 mm. It is driven by a built-in single-phase electrical motor (250 V) and available as both a table and floor model. The meat is put onto the tray and continuously fed by hand into a vertical cylindrical hole leading to the feed auger. The meat or fat is forced by its own weight into the barrel with the rotating feed auger. This type of meat grinder is the most suitable for commercial small-scale operations. Some brands use the Enterprise cutting system, others the Unger system.

Large industrial meat grinders are driven by a three-phase electrical motor (400 V) and equipped with the Unger cutting system. The orifice cylinder diameter of this type of grinder ranges from 114 - 400 mm. Industrial grinders are either designed as stuffing grinders with either tray or hopper or as an automatic mixing grinder. The automatic mixing grinder has a big hopper and the meat falls automatically onto the mixing blades and the feeding worm (auger). The mixing blades and feeding worm can be operated independently with mixing blades rotating in both directions but the feeding worm only towards the cutting set. Most of the industrial meat grinders are also equipped with a device for separating tendons, bone particles and cartilage.

Bowl Cutter

The bowl cutter is the commonly used meat chopping equipment designed to produce small or very small lean meat and fat particles. Bowl cutters consist of a horizontally revolving bowl and a set of curved knives rotating vertically on a horizontal axle at high speeds of up to 5,000 rpm. Many types and sizes exist with bowl volumes ranging from 10 to 2000 litres. The most useful size for small- to medium-size processing is 20 to 60 litres. In bigger models bowl and knife speed can be regulated by changing gears. Bowl cutters are equipped with a strong cover. This lid protects against accidents and its design plays a crucial role in the efficiency of the chopping process by routing the mixture flow. Number, shape, arrangement, and speed of knives are the main factors determining the performance of the cutter. Bowl cutters should be equipped with a thermometer displaying the temperature of the meat mixture in the bowl during chopping.

Modern large scale bowl cutters may have devices to operate under a vacuum, which helps to improve colour and texture of the meat products by keeping oxygen out of the meat mixes and avoid air pockets. Cutter knives should be adjusted to a distance of 1-2 mm from the bowl for optimal cutting (check the manufacturers recommendations for each model). Most of the large and high-speed bowl cutters are equipped with mechanical discharger devices for emptying the cutter. The process of chopping in a bowl cutter is used for producing fine comminuted products such as frankfurters, bologna, liver sausage etc., and enables processors to offer a much wider range of products.

Filling Machines

These machines are used for filling all types of meat batter in containers such as casings, glass jars, cans etc. The most common type of filling machine in small and medium size operations is the piston type. A piston is moved inside a cylinder forcing the meat material through the filling nozzle (funnel, stuffing horn) into the containers. Piston stuffers are either attached to the filling table or designed as floor models. In small-scale operations manual stuffers are usually sufficient, sometimes even simple hand-held funnels are used to push meat mixes into casings.

Clipping Machines

Clipping machines are used for artificial or natural casings. Clipping machines can also be connected to filling machines. Such machines work with so called casing brakes, which are devices for slow release of the shirred casings from the filling horns ensuring tight filling. Then the filled casing segments are clipped in portions. So called double clipping machines place two clips next to each other, which ensures that the individual sausage portions remain clipped on both ends and easy separation of the sausage portions is possible. When using shirred casings, the time consuming loading of pre-cut casings is no longer necessary. Wastage of casings can be reduced to a minimum by tight filling and leaving only as much casing for the sausage end as needed for the placing of the clips.

Clipping machines are mainly used in larger operations and in most cases operated by compressed air. For medium-scale operations manually operated hand clippers are available.

Smokehouses

Simple smokehouses are used for smoking only. In traditional and small-scale operations the most common methods of smoke generation include burning damp hardwood sawdust, heating dry sawdust or heating pieces of log. But technological progress has changed the smoke generation and application techniques. Methods used in modern meat processing include the following:

Burning/smouldering of saw dust

In modern smokehouses (1), smoke generation takes place outside the smoking chamber in special smoke generators with electrical or gas ignition (4). Separate smoke generators allow better control of the quantity and temperature of the smoke produced. The sawdust or chip material (3) is moved from the receptacle to the burning zone (4) by a stirrer or shaker (3). It is ignited by means of an electrically heated plate or by gas flame. A smoke stripper, which is basically a cold water spray, can be placed in the initial part of the smoke pipe and serves to increase the purity of the smoke as undesirable substances are washed out. Smoke with a high degree of desirable smoke components can be obtained in the low temperature range of thermal destruction of saw dust beginning at around 230°C and not exceeding 400°C.

Smoke generation through friction

Timber (3), which is pressed (1) against a fast-rotating steel drum (4) results in pyrolysis of the wood in the favourable temperature range of 300°C to 400°C. The flameless, light, dense and aromatic smoke contains a large proportion of desirable smoking substances and a low proportion of tars. The smoke is conveyed (2) into the smoking chamber. The creation of smoke can be commenced and completed in a matter of seconds. The operation of this type of smoke generators is usually carried out in a discontinuous manner. The smoke quantity and quality can be regulated by changing the speed and time of rotation. As this type of smoke can be produced at relatively low temperatures, it does not carry high amounts of hazardous substances such as benzopyrene.

Smoke generation through steam

Overheated steam (3) at approximately 300°C is injected into a compact layer of sawdust (4), which causes thermal destruction of the wood and smoke is generated. This method allows the control of smoke generation temperature by choosing the adequate steam temperature. Impurities in the smoke caused by particles of tar or ash are minimal. The steam-smoke mixture condensates extremely quickly and intensively on the surface and inside the sausage products and produces the desired smoking colour and flavour. No connection to the chimney is required as smoke particles not entering the products settle down in the condensing steam. The condensed water is conducted to the effluent system.

Combined Equipment

Modern facilities can combine smoking, cooking and cooling operations for meat products in one continuous process. By means of automatic stirring systems processing parameters such as smoke generation, temperature and relative humidity required to dry, smoke, or steam-cook any type of product, can be pre-set. With additional refrigerated units installed in the smokehouse, it is also possible to use it as a fermenting/ripening room for the first crucial steps in production of fermented sausages or raw ham products, where air temperature and air humidity have to be accurately controlled.

Brine Injector

This equipment serves for the injection of brine into meat. Brine is water containing dissolved salt and curing substances (nitrite) as well as additives such as phosphates, spices, sugar, carrageenan and/or soy proteins. The injection is done by introducing pointed needles into the muscle tissue. Brine injection is mainly used for the various types of ham, bacon and other whole muscle products.

Brine injectors are available in different sizes from manually operated single-needle devices for small-scale operations to semi-automated brine injectors with up to 32 needles and more. In large machines the quantity of brine injected into the fresh meat can be determined by pre-setting of pressure and speed. It is very important that all parts of the brine injectors are thoroughly cleaned after every working session and disinfected regularly.

Before the injector is used again all hoses and needles should be rinsed with warm water as particles left in the system can block the needles. Absolute cleanliness is necessary as microorganisms remaining in the system would be injected deep into the meat pieces during the operation.

Tumbler or Massager

Tumblers are used for the processing of meat products such as whole-muscle or reconstituted hams. Such machines resemble in principle a drum concrete mixer. A rotating drum with steel paddles inside slowly moves the meat pieces thus causing a mechanical massaging effect. This mechanical process is assisted by the addition of salt and phosphates to achieve equal brine distribution and liberates muscular protein from the meat tissue (protein extraction). The semi-liquid protein substances join the meat pieces firmly together during later heat treatment. For hygienic reasons it is important to place the tumbler below 10°C to avoid excessive microbial growth during lengthy tumbling times (more then 4 hours or even over night). In specific cases it is recommended that the tumbler should be operated refrigerated or inside a cold room below -1°C, as these temperatures are best to extract as much soluble protein as possible from the muscle meat.

Vacuum Packaging Machine

For vacuum packaging the meat product has to be placed into a vacuum bag. Air is removed from the bag by means of the vacuum packaging machine and the bag then sealed. Special vacuum packaging machines can operate with so called gas-flushing, where a mixture of gas is injected after evacuating the air. Such protective gas atmospheres inside the product package inhibit bacterial growth and stabilize the meat colour. The gas mixtures usually contain CO_2 and N_2.

Mixer/Blender

Mixers are used to blend meat and spices, or coarse and finely chopped meat. The machine generally consists of a rectangular or round bottom vessel through which two parallel shafts operate. Various paddles are mounted on those shafts to mix the meat. The mixer is discharged through tilting by 90 degrees. Some mixers are designed as vacuum mixers, as the mixing under vacuum (exclusion of oxygen) has advantages for the development of desirable product colour and texture.

The emulsifier serves for the preparation of very fine meat emulsions. Its functional parts are a perforated plate, attached to which two edged blades are rotating. Next to the blades there is a centrifugal pump that forces the pre-ground meat through the perforated plate. Most emulsifiers are vertical units. Compared to the bowl cutter the emulsifier operates at much higher speed, producing a finer emulsion-like mix. The emulsifier is also perfectly suited to produce semi-processed products such as pig skin emulsions.

Ice Flaker

In these machines ice flakes are continuously produced from potable water. Ice is needed in meat processing for some types of meat products. Water, added in the form of ice, is an important ingredient in order to enhance protein solution and to keep the temperature of the meat batter low. Ice flakers with in-built UV-water-disinfection device are available for areas with unsafe water supply.

Frozen Meat Cutter

The purpose of cutting frozen meat blocks into smaller pieces is to make frozen meat suitable for immediate comminuting in grinders, bowl cutters etc. without previous thawing. There are two types of machines for the cutting of frozen meat blocks, working either with knives cutting in vertical direction or using rotating drums with attached sharp knives. In the guillotine-type machines a knife head is driven hydraulically and even the hardest frozen products can be cut into small pieces, either meat cubes or meat strips. Rotary frozen meat cutters operate according to the principle of carving out particles from the frozen meat blocks. The rotary drums can be equipped with knives capable of cutting out pieces of frozen meat from large fist-size to small chip-size.

Meat Processing–Standard Practices

Meat processing technologies include on the one hand purely technical processes such as

- Cutting, chopping, comminuting
- Mixing, tumbling
- Stuffing/filling of semi-fabricated meat mixes into casings, synthetic films, cans etc.
- Heat treatment

On the other hand, chemical or biochemical processes, which often go together with the technical processes, are also part of meat processing technology such as

— Salting and curing
— Utilization of spices and additives
— Smoking
— Fermentation and drying

Cutting

There are five methods of mechanical meat cutting for which specialized machinery is used:

Mincing (grinding) of lean and fatty animal tissues

Larger pieces of soft edible animal tissues can be reduced in size by passing them through meat grinders. Some specially designed grinders can also cut frozen meat, others are equipped with devices to separate "hard" tissues such as tendons and bone particles from the "soft" tissues.

Chopping animal tissues in bowl cutter (discontinuous process)

Bowl cutters are used to chop and mix fresh or frozen lean meat, fat (and/or edible offal, if required) together with water (often used in form of ice), functional ingredients (salt, curing agents, additives) and extenders (fillers and/or binders)

Chopping animal tissues in emulsifying machines (continuous process)

The animal tissues to be emulsified must be pre-mixed with all other raw materials, functional ingredients and seasonings and pre-cut using grinders or bowl cutters. Thereafter they are passed through emulsifiers (also called colloid mills) in order to achieve the desired build-up of a very finely chopped or emulsified meat mix.

Frozen meat cutting

Boneless frozen meat blocks can be cut in slices, cubes or flakes by frozen meat cutters or flakers. The frozen meat particles (2-10 cm) can be directly chopped in bowl cutters without previous thawing thus avoiding drip losses, bacterial growth and discoloration which would happen during thawing. For small operations the manual cutting of frozen meat using cleavers or axes is also possible.

Cutting of fatty tissues

Back fat is cut in cubes of 2-4 cm on specialized machines to facilitate the subsequent chopping in cutters/emulsifiers. In small-scale operations this process can be done manually.

Salting/Curing

Salting

Salt (sodium chloride NaCl) adds to the taste of the final product. The content of salt in sausages, hams, corned beef and similar products is normally 1.5-3%. Solely common salt is used if the cooked products shall have a greyish or greyish-brown colour as for example steaks, meat balls or "white" sausages. Besides adding to flavour and taste, salt also is an important functional ingredient in the meat industry, which assists in the extraction of soluble muscle proteins. This property is used for water binding and texture formation in certain meat products.

The preservation effect, which is microbial inhibition and extension of the shelf-life of meat products by salt in its concentrations used for food (on average 1.5-3% salt), is low. Meat processors should not rely too much on this effect unless it is combined with other preservation methods such as reduction of moisture or heat treatment.

Curing

Consumers associate the majority of processed meat products like hams, bacon, and most sausages with an attractive pink or red colour after heat treatment. However experience shows that meat or meat mixes, after kitchen-style cooking or frying, turn brownish-grey or grey. In order to achieve the desired red or pink colour, meat or meat mixes are salted with common salt (sodium chloride NaCl), which contains a small quantity of the curing agent sodium nitrite ($NaNO_2$). Sodium nitrite has the ability to react with the red meat pigment to form the heat stable red curing colour. Only very small amounts of the nitrite are needed for this purpose.

Nitrite can be safely used in tiny concentrations for food preservation and colouring purposes. Traces of nitrite are not poisonous. In addition to the reddening effect, they have a number of additional beneficial impacts so that the meat industries widely depend on this substance. Levels of 150 mg/kg in the meat product, which is 0.015%, are normally sufficient.

To reduce the risk of overdosing of nitrite salt, a safe approach is to make nitrite available only in a homogeneous mixture with common salt generally in the proportion 0.5% nitrite and the balance of sodium chloride (99.5%). This mixture is called nitrite curing salt. At a common dosage level of 1.5-3% added to the meat product, the desired salty flavour is achieved and at the same time the small amount of nitrite needed for the curing reaction is also provided. Due to the sensory limits of salt addition (salt contents of 4% are normally not exceeded), the amounts of nitrite are kept low accordingly.

A great deal of research has been done with regard to the utilization of nitrite and it can be said that nitrite in meat products is safe if basic rules are adhered to. Nitrite is now recognized a substance with multifunctional beneficial properties in meat processing:

— The primary purpose of nitrite is to create a heat resistant red colour in a chemical reaction with the muscle pigment, which makes cured meat products attractive for consumers.

— Nitrite has a certain inhibitory effect on the growth of bacteria. This effect is particularly pronounced in canned meat products which are usually stored without refrigeration, where small numbers of heat resistant bacteria may have survived but their growth is inhibited by the presence of nitrite.

— Nitrite has the potential of attributing a specific desirable curing flavour to cured products.

— In the presence of nitrite fats are stabilized and rancidity in meat products retarded i.e., an antioxidant effect.

Many attempts have been made to replace nitrite by other substances, which would bring about the same beneficial effects as listed above. Up to now no alternative substance has been found. As the above desirable effects are achieved with extremely low levels of nitrite, the substance can be considered safe from the health point of view. Currently the known advantages of nitrite outweigh the known risks.

Curing of chopped/comminuted meat mixtures

Curing is applied for most chopped meat mixtures or sausage mixes for which a reddish colour is desired. The curing agent nitrite is added in dry form as nitrite curing salt. The reaction of nitrite with the red meat pigment starts immediately. Due to homogenous blending the meat pigments have instant contact with the nitrite. Higher temperatures during processing, e.g. “reddening” of raw-cooked type sausages at 50°C or scalding/cooking of other products at 70-80°C, accelerate the process.

Another accelerating or “catalytic” effect is the addition of ascorbic acid, which slightly lowers the pH of the meat mixture. However, the dosage of ascorbic acid must be low (0.05%), just to provide the slightly acid conditions for the reduction of $NaNO_2$ to NO. A pronounced reduction of the pH would negatively affect the water binding capacity of the product which is not desirable.

Curing of entire meat pieces

Besides the curing of chopped meat mixtures, entire pieces of muscle meat can be cured. However, due to size the curing substances cannot instantly react with the meat pigments as is the case in chopped meat mixes. Hence various curing techniques are applied.

The final products of curing entire meat pieces are either cured raw fermented products or cured cooked products. The curing system to be used depends on the nature of the final product (uncooked or cooked). There are two systems for curing entire meat pieces, dry curing and wet curing (“pickling”) and the type of the final product determines which system will used.

In dry curing a curing mix is prepared containing salt or nitrite curing salt, together with spices and other additives. The pieces of meat are rubbed with this curing mix and packed in tanks. The curing mix gradually permeates into the meat, which can be a lengthy process ranging from several days to several weeks.

The second method of curing meat pieces is wet curing, also called pickling, which involves the application of curing brine to the meat. For the manufacture of the brine, curing salt and spices, and other additives if required are dissolved in water. The meat cuts are packed in tanks and brine is added until all pieces are completely covered. A temperature of +8 to +10°C for the curing room is recommended as lower temperatures may retard curing. For equal penetration of the brine, the meat is cured for periods ranging from several days to two weeks depending on the size of the cuts and curing conditions. After completion of the curing, ripening periods for the products follow for taste and flavour build-up.

An alternative and quick way of wet curing is to accelerate the diffusion of the curing substances by pumping brine into the meat tissue ("injection curing"). For this purpose brine injectors with perforated hollow needles are used. The injection of brine into the muscles can be done manually by using simple pumping devices. At the industrial level semi-automatic multi-needle brine injectors are used which achieve very even distribution of the curing ingredients and can reduce the curing period (equal distribution of the curing substances or "'resting period") to less than 48 hours.

In addition, most injection cured meat pieces which are to be processed into cured-cooked products (such as cooked hams etc), are submitted to a tumbling process. Tumbling further accelerates the brine penetration throughout the meat prices and "resting periods" are not necessary.

Smoking

Smoke for treatment of meat products is produced from raw wood. Smoke is generated through the thermal destruction of the wood components lignin and cellulose. The thermal destruction sets free more than 1000 desirable or undesirable firm, liquid or gaseous components of wood.

These useful components contribute to the development of the following desirable effects on processed meat products:

— Meat preservation through aldehydes, phenols and acids (anti-microbial effect)
— Antioxidant impact through phenols and aldehydes (retarding fat oxidation)
— Smoke flavour through phenols, carbonyls and others (smoking taste)
— Smoke colour formation through carbonyls and aldehydes (attractive colour)
— Surface hardening of sausages/casings through aldehydes (in particular for more rigid structure of the casing)

The most known undesirable effect of smoking is the risk of residues of benzopyrene in smoked products which can be carcinogenic if the intake is in high doses over long periods. With normal eating habits, a carcinogenic risk is normally not associated with moderately smoked food such as smoked meat products.

Depending on the product, smoke is applied at different temperatures. There are two principal smoking techniques:

— Cold smoking

— Hot smoking

The principle of both methods is that the smoke infiltrates the outside layers of the product in order to develop flavour, colour and a certain preservation effect.

Cold Smoking–This is the traditional way of smoking of meat products and was primarily used for meat preservation. Nowadays it serves more for flavour and colour formation, for example in sausages made from precooked materials such as liver sausage and blood sausage.

The combination of cold smoking and drying/ripening can be applied to fermented sausages and salted or cured entire meat pieces, in particular many raw ham products. In long-term ripened and dried hams, apart from providing colour and favour, the cold smoking has an important preservative effect as it prevents the growth of moulds on the meat surfaces.

The optimal temperature in "cold" smoking is 15 to 18°C (up to 26°C). Sawdust should be burned slowly with light smoke only and the meat hung not too close to the source of the smoke. Cold smoking is a long process which may take several days. It is not applied continuously, but in intervals of a few hours per day.

Hot Smoking–Hot smoking is carried out at temperatures of +60 to 80°C. The thermal destruction of the wood used for the smoking is normally not sufficient to produce these temperatures in the smoking chamber. Hence, additional heat has to be applied in the smoking chamber.

The relatively high temperatures in hot smoking assure a rapid colour and flavour development. The treatment period is kept relatively short in order to avoid excessive impact of the smoke.

Hot smoking periods vary from not much longer than 10 minutes for sausages with a thin calibre such as frankfurters to up to one hour for sausages with a thick calibre such as bologna and ham sausage and products like bacon and cooked hams.

Products and smoking–Cold smoking is used for fermented meat products (raw-cured ham, raw-fermented sausage) and precooked-cooked sausage (liver and blood sausages). Hot smoking is used for a range of raw-cooked sausages, bacon and cooked ham products. Smoke treatment can only be applied, if meat the products are filled in casings permeable to smoke. All natural casings are smoke permeable, as are cellulose or collagen basis synthetic casings.

Smoke permeable casings can also be treated using a new technology, where a liquid smoke solution is applied on the surface. This can be done by dipping in solution, showering (outside chamber) or atomization (spraying inside chamber). Polyamide or polyester based synthetic casings are not permeable to smoke. If smoke flavour is wanted for products in such casings, small quantities of suitable smoke flavour (dry or liquid) are added directly to the product mix during manufacture.

Selection and Grading of Raw Materials

The *two main components of processed meat products* are *animal muscle meat* and *animal fat*. Apart from pure muscle tissue,muscle meat also contains some connective tissue and inter- and intra-muscular fat, which determine the quality of muscle meat. Animal fats are of firmer or softer texture depending on their location in the animal body. In addition to the animal species, the texture of fats determines their processing quality. Edible animal by-products such as skin, internal organs and blood also play a role as raw materials for meat processing. By-products are not generally used; they are part of specific processed meat products.

The first preparatory step for processing of meat into meat products is the product-oriented selection of raw animal materials, taking into account their quality and processing suitability and the characteristics of the meat products to be fabricated. Some meat products require lean meat without adhering fat or connective tissue, while others have a higher fat and/or connective tissue contents. Other products require firm animal fats, for others soft fats are better suited. Choosing appropriate raw materials is indispensable for efficient meat processing and is best done by visual selection and grading according to the tissue-specific properties.

Meat processors are advised to develop enterprise-specific standards of raw material composition for each meat product fabricated. The proper grading of raw materials, which needs skills and experience, has a decisive impact on the quality of the meat products and resulting revenues which can be generated.

For the needs of small to medium sized meat processing plants, simple grading schemes are described hereunder, with raw materials from pigs, cattle/buffaloes and other ruminants as well as poultry.

Non-meat Ingredients

Along with the main components *meat* and animal *fat*, a wide range of substances of *non-meat origin* are used as ingredients in processed meat products. Some of them are absolutely necessary, such as salt and spices. Others are used for specific products.

One way of categorizing non-meat ingredients is by source. They are either

— chemical substances or

— of plant origin or

— of animal origin.

Other classification criteria for non-meat ingredients are, whether they are *additives* or *full foods* or whether they have *functional* properties or not.

Additives are usually substances, which are not normally consumed as food by itself, but which are added to develop certain technological and quality characteristics. In contrast, vegetables, flours, eggs, etc. could be considered as full food ingredients.

Most ingredients are functional, which describes their ability to introduce or improve certain quality characteristics. The functional properties of ingredients include their impact on:

— taste

— flavour

— appearance

— colour

— texture

— water binding

— counteracting fat separation

— preservation

Ingredients which are solely functional without any other effect such as filling or extending the volume of the product, are normally used in small amounts.

Meat loaf cut, left with curing colour, centre without colour, right with artificial colour

The criteria for the utilization of functional non-meat ingredients are:

— safe for consumers, and

— improve of processing technology and/or sensory quality of the products.

In contrast to the exclusively functional substances, there is another group of ingredients that are not primarily intended for change of appearance or quality improvements but serve to *add volume* to the meat products. They are called meat extenders and fillers. Their main purpose is to make meat products *lower-cost*. Meat extenders and fillers include cereals, legumes, vegetable, roots and tubers and are used in larger quantities, on average between 2 and 15%.

Meat extenders are primarily plant *proteins* from legumes, with *soybeans* as the major source. *TVP* (Textured Vegetable Protein) is the most common soy bean extender. These cheaper plant proteins *"extend"* the more expensive meat proteins, resulting in acceptable overall protein contents of lower cost meat products. Extenders are added in sizeable amounts that increase the bulk of the meat products, but this may also alter their quality. From animal protein sources, whole milk and eggs can be considered as meat extenders. In some countries, replacement of meat by fish is gaining popularity resulting in fish products which may be meat and fish mixes or entirely made of fish materials, e.g. "fish viennas", made using the same technology and process as viennas made of meat.

Fillers are also mostly plant substances, low in protein and high in *carbohydrates* such as cereals, roots, tubers and vegetables and some refined products such as *starches* and *flours*. Pure meat products are very low in carbohydrates. Hence the addition of carbohydrate-rich substances is not an "extension" of the protein mix, but some new components *"fill-up"* the product volume. Apart from their volume-filling capacity, some fillers, in particular starches and flours, are also used for their capability to absorb extensive quantities of water.

Extenders and fillers are *not* standard ingredients in processed meats, in fact high quality products are often manufactured without them. But they are useful tools in *cost reduction* enabling the manufacture of lower-cost but still nutritive meat products. Such products are particularly suitable to supply valuable animal proteins in the diets of consumers who cannot regularly afford expensive meat and meat preparations.

As another definition for specific non-meat ingredients, the term binder is used for substances of animal or plant origin, which have a significant high level of protein that serves for both *water and fat binding*. Such substances include *high-protein soy, wheat andmilk products*, such as soy isolate, wheat gluten, milk protein (caseinate). They are not extenders in the first place due to the low quantities added (approx. 2%), but act through their high quality proteins that are instrumental in water binding and protein network structuring. On the other hand, some substances with little or no protein level, like *starches* and *flours* mentioned above under "fillers", can bind water and fat by means of physical entrapment and could also be considered "binders".

The above aspect illustrates that clear definitions in the wide range of non-meat ingredients are difficult to establish. While most substances have one *dominating effect*, there are in many cases also desirable *side effects* that, however, complicate their clear grouping. Even those substances like textured vegetable protein/TVP, which are primarily intended for non-functional purposes, namely meat extension, have a water binding effect, which qualifies them also as moderately functional. Also soy isolates or dried milk powders, which are used as binders, also have a slight extension effect as the amounts added moderately increases the protein level. Most substances have double or even multiple effects.

Therefore, in order to provide an overview of the most common substances used as non-meat ingredients, they are listed hereunder according to their origin, namely *chemical* (a) or of *animal* (b) or *plant origin* (c):

Chemical Substances Used as Ingredients

There are various chemical substances approved for the different kinds of food processing, but in the specific case of meat processing the number of approved chemical substances is rather limited in most countries. The following are of significance:

— *Salt* (for taste, impact on meat proteins, shelf-life)

— *Nitrite* (for curing colour, flavour, shelf-life)

— *Ascorbic acid* (to accelerate curing reaction)

— *Phosphates* (for protein structuring and water binding)

— *Chemical preservatives* (for shelf-life)
— *Antioxidants* (for flavour and shelf-life)
— *Monosodium glutamate MSG* (for enhancement of flavour)
— *Food colouring substances* (synthetic and of plant origin)

Chemical additives have exclusively functional properties, they are used in small amounts usually below 1% (with nitrate as low as 0.05%). Only salt is in the range of 2% (with up to 4% in some fermented dried products).

Non-meat Ingredients of Animal Origin

Ingredients of animal origin are not commonly applied but may be useful for specific meat preparations. They all have functional properties (except whole milk), in particular improvement of water binding and prevention of fat separation during heat treatment. Apart from their functional properties, some of them can also be considered meat extenders, as mentioned below.

— *Milk caseinate* (90% protein; used in small quantities (2%); have functional water and fat binding properties)
— *Whole milk* or *non-fat dried milk* (=skim milk) (sometimes used in indigenous meat preparations as a protein extender)
— *Gelatine* (binding properties and meat extender)
— *Blood plasma* (predominantly binding properties)
— *Eggs* (extender and binding ingredient for meat pieces and fried sausages)
— *Transglutaminase* (exclusively binding properties)

Ingredients of Plant Origin

All *spices* are of plant origin. They are predominantly *functional* and used in small quantities to provide or add flavour and taste to meat products.

Another group of predominantly *functional* substances of plant origin with high protein content are used as *binders* to increase water binding and fat retention, in particular in intensively heat treated products. The most commonly used substances are

— *isolated soy protein* (90% protein) and
— *wheat gluten* (80% protein)
— and, less importantly, protein isolates from other legumes.

A third group of ingredients of plant origin are used as *meat extenders* (if rich in proteins) or *fillers* (if rich in carbohydrates) for meat product and sausage formulations. The purpose is to replace expensive meat for lower- or medium-grade products by cheaper ingredients of plant origin for cost reduction and volume increase.

Application of non-meat ingredients

For the application of ingredients listed above to meat products, various methods are deployed, depending upon the properties of the ingredient and the meat product. A uniform distribution is crucial for equal intensity of flavour, colour, texture or any other quality characteristic expected from the product.

Methods of application

a) *During grinding.* Chemical additives and smaller quantities of other fine or coarse non-meat ingredients or granulated substances (such as TVP) are easily incorporated in ground meat products by mixing them with the raw meat materials prior to grinding. In small scale operations, the mix of meat and non-meat ingredients is then simply passed through the grinder plates. Manual or mechanical blending can be added if necessary. In larger industrial operations and for heavily extended products, ground meat materials, chemical additives and other non-meat ingredients are usually combined in a blender.

b) *During chopping.* In finely comminuted or chopped meats, non-meat ingredients are easily dispersed by mixing them with the rest of the batter in comminuting equipment. Non-meat ingredients such as binders (isolated soy protein/ISP, milk caseinate) are preferably added in emulsion form, finely milled fillers (flours, starches) in dry form. In smaller calibre low-cost sausages such as hotdogs, also larger quantities of extenders (e.g. re-hydrated TVP) and coarse fillers (rusk, breadcrumbs, etc) are incorporated during the chopping process.

c) *Application to non-comminuted meat.* The addition of non-meat ingredients to larger meat pieces or intact muscles is more complex. Injection of ingredients as part of the curing brine, if they are water soluble or can be dispersed in water (salt, nitrite, spices, ascorbate, phosphates, soy products, Carrageenan), is the most rapid method of equal distribution. The surface application of such dry substances (e.g. nitrite curing salt, spices) or immersion of meat in salt/curing salt and flavouring solutions is another way of application, but requires days or weeks to diffuse throughout the muscle tissue.

Seasonings Used in Meat Processing

Seasonings are normally parts of plants which flavour food. The trade in and the processing of spices has developed into an important support industry for food processing enterprises in order to meet consumer preferences. Mixtures of seasonings were developed in order to serve as flavouring agents for various meat products. *Natural spices, herbs and vegetable* bulbs are the main groups of seasonings and are described hereunder.

Natural Spices

The term "natural spices" includes dried rootstocks, barks, flowers or their parts and fruits or seeds of different plants. The most important natural spices used in processed meat products are pepper, paprika, nutmeg, mace, cloves, ginger, cinnamon,cardamom, chilli, coriander, cumin and pimento. The most common natural spice in sausage making is pepper. Spices are mainly used in the ground form with particle sizes from 0.1 to 1 mm.

Herbs

Herbs are dried leaves of plants grown in temperate climates. The major herbs used in processed meat products are basil, celery,marjoram, oregano, rosemary and thyme.

Vegetable Bulbs

The main natural seasonings originating from vegetable bulbs and used in processed meat products are onions and garlic.

Extracts

Natural spices are often contaminated with high numbers of microorganisms, in particular spores, due to their production process. This may become a problem for the stability of the meat products. The microbial load of spices can be reduced by irradiation orfumigation. Such treatments are not allowed everywhere. Another option is the use of spices extracts. Extracts are produced by separating the flavour-intensive fractions through physico-chemical procedures (e.g. steam distillation) which results in germ-free flavouring substances. Extracts are preferably used in viscous liquid or oily form. Due to the absence of microorganisms, extracts are specifically recommended for the production of microbiologically sensitive processed meat products, such as cured-cooked hams or cured-cooked beef cuts.

Heat Treatment of Meat Products

Heat treatment of processed meat products serves two main purposes:

— Enhancement of desirable texture, flavour and colour, in order to make meat products more palatable and appetizing for consumption.

— Reduction of microbial content thus achieving the necessary
  - preservation effects for an extended shelf life (storability) of the products and
  - food safety effects by eliminating potential food poisoning agents

The heating parameters to be applied in meat processing can vary considerably in temperature and time depending on the type of product. Heat treatment methods cause various physical-chemical alterations in meat, which result in the beneficial sensory andhygienic effects on the processed products.

When mankind learned to use fire for food preparation, the aspects of palatability were clearly important. Heat treatment became the common way of making meat palatable for consumption. The impact of high temperatures induces coagulation and denaturation of meat proteins and structural and chemical changes of fats and carbohydrates, which make meat tastier and also more tender. In addition, the absorption of nutrients from heat treated meats in the digestive tract of humans is improved.

In modern times, with longer distribution channels for meat and the popularity and steadily growing quantities of processed meat products on the markets, the hygienic aspects of heat treatment of such processed meats, which result in germ reduction, became increasingly important.

Heat Treatment for Microbial Control

Contrary to meat dishes, which are usually consumed hot immediately after preparation, most processed meat products are heat treated during manufacture and cooled down in a next step, as they undergo shorter or longer cold storage periods for distribution and sales. Hence, processed products must have an adequate shelf life, which can only be achieved if their microorganism content is low or practically zero. During slaughtering, subsequent meat cutting and initial processing steps, the numbers of microorganisms in meat are steadily increasing. The thermal treatment at the end of the processing stage is therefore important for microbial control. It is the effective tool to reduce or eliminate the contaminating microflora.

Heating Parameters for Meat Products

For preparation of meat dishes in households or restaurants, exact temperature control is normally not needed and it is only differentiated between low, medium and high dry or moist heat. Meat dishes are usually consumed immediately after cooking, so the heat treatment is (besides basic food safety aspects) mainly for sensory reasons. The achievement of a prolonged shelf life is not intended.

For processed meat products exact temperature control is indispensable, as the balance between two opposite requirements has to be found:

— Heat treatment temperatures should be raised high enough to accomplish adequate microbial reduction for shelf life extension.

— Heat treatment temperatures should be kept low enough to prevent deterioration of the eating quality.

Heat treatment of processed meat products will therefore always be a compromise between sensory and hygienic requirements.

In case of difficult hygienic conditions (e.g. tropical environment, highly contaminated raw meat, risk of interrupted cold chain) more intensive heat treatment must be applied. However, this may result in a certain degradation of the eating quality and higher cooking losses. If meat production and meat handling conditions are good (e.g. moderate climate, fresh hygienic raw materials, excellent processing and storage conditions), the heat treatment can be less intensive, which results in better sensory quality, but in hygienically more sensitive products.

Hurdle Technology of Heat-treated Products

In modern meat processing, the effect of heat treatment can be supported by the application of additional "hurdles", which have the potential to slow down microbial growth. Such "hurdles" allow keeping the heat treatment of sterilized products at lower temperature levels, so that the product quality is less affected. Alternatively, this technology can be used to produce shelf-stable products of the non-sterilized type through heat treatments below 100°C. This kind of heat treatment alone would not be enough to stop microbial growth, but the additional "hurdles" complete the effect. This kind of meat preservation is called hurdle technology.

Frequently used "hurdles" are lowering of water activity (a_w) or acidity (pH) in a product, or the utilization of chemical preservatives, to which amongst many others also the commonly used nitrite curing salt belongs. All these measures on their own would not stop microbial growth, but some or all of them in combination with heat treatment account for a number of "hurdles", which cannot be overcome by microorganisms surviving in the product. The result of such "built-in hurdles" is that meat products can be moderately heated, but surviving microorganisms can not grow. In most efficient combinations of such "hurdles", microorganisms do not even grow under ambient ("room temperature") storage conditions. Such products do not need refrigeration, they are shelf-stable, but much less heat treatment was needed than for fully sterilized canned products. Naturally, in the meat sector the range of products, which can be made shelf-stable according to the hurdle technology, is limited but may be of significance in certain circumstances, in particular if no uninterrupted cold chain is available.

Types of Heat Treatment

Principally, for heat treatment (also called "thermal treatment") of meat and meat products, it can be distinguished between products which undergo

a. Heat treatment at temperatures below 100°C, mostly in the temperature range of 60 to 85°C, also called "pasteurization" or simply "cooking".

b. Heat treatment at temperatures of above 100°C, also called "sterilization".

All such products will achieve a more or less prolonged shelf life through reduction or complete destruction of microbial populations by the heating process (thermal reduction/thermal destruction).

Categories of Processed Meat Products

When viewing meat products of various size, shape and colour in butcher shops or meat sections of supermarkets, there appears to be is a great variety of such products with different taste characteristics. In some countries there may be several hundred different meat products, each with its individual product name and taste characteristics.

At a closer look, however, it turns out that many of the different products with different product names have great similarities. This issue can be even better understood and becomes more transparent when the processing technologies are analyzed. Based on theprocessing technologies used and taking into account the treatment of raw materials and the individual processing steps, it is possible to categorize processed meat products in six broad groups.

Fresh Processed Meat Products

These products are meat mixes composed of comminuted muscle meat, with varying quantities of animal fat. Products are salted only, curing is not practiced. Non-meat ingredients are added in smaller quantities for improvement of flavour and binding, in low-cost versions larger quantities are added for volume extension. All meat and non-meat ingredients are added fresh. Heat treatment is applied immediately prior to consumption to make the products palatable. If the fresh meat mixes are filled in casings, they are defined as sausages. If other portioning is customary, the products are known as patties, kebab, etc. Convenience products, such as chicken nuggets, have a similar processing technology and can also be included in this group. In contrast to the rest of the group, chicken nuggets etc. are already fried in oil at the manufacturing stage during the last step of production.

This group comprises meat mixes composed of finely comminuted, minced or sliced muscle meat, with varying quantities ofanimal fat adhering to the muscle meat or added separately. Flavouring is done by adding common salt and spices; curing is not practiced. In many products other non-meat ingredients are added in smaller quantities for improvement of flavour and binding, in low-cost versions larger quantities are added to extend the existing volume. The characteristic of this group is that all meat and non-meat ingredients are added fresh (raw), either refrigerated or non-refrigerated. The heat treatment (frying, cooking) is only applied immediately prior to consumption to make the products palatable. In many instances, the consumer cooks the products prior to serving and products are consumed hot. Most of the fresh meat mixes are filled in casings, which defines such products as sausages. If other portioning is customary, the products are known as burgers, patties, kebab, etc.

Patties, Kebab, etc.

Patties are formed from minced meat usually in a disc-like shape with diameters of 80-150mm and 5-20mm height. In commercial fast-food outlets the common name is hamburgers or simply burgers. Originally, burgers were made from *beef*(preferably lean cow meat), but in recent years *chicken* and *mutton* burgers have become more common. Other animal tissues such as fats or connective tissue/tendons can also be part of the mixture, with quantities depending on the type and quality of the products. In industrial manufacture, these tissues could have been previously separated from the lean meat and are added again in defined quantities to ensure identical chemical composition (protein, fat, water) of all products. A common feature of burgers is that during mincing (1-3mm disc) and consecutive blending, salt and spices (mainly black and white pepper, in some instances also herbs, garlic or onions) are added. In some cheaper industrial formulations textured soy protein is commonly used as a non-meat ingredient in quantities up to 25%. Other non-meat ingredients suitable for this purpose could include rusk, breadcrumbs and dried flakes from roots and tubers.

Burgers are stored frozen and individually pan-fried before consumption. Ideally, internal temperatures of 80°C should be reached to destroy food poisoning agents potentially present in the raw meat mixes. Burgers are often served on bread rolls or buns with slices of cheese, mayonnaise, mustard, green salad, etc.

The Kebab is a Middle East product, but popular in many places and usually eaten in pieces of flat white bread with yogurt sauce or sheep cheese. These preparations of kebab are also known by the name of doener or gyros. The term "kebab" refers to processed meat on skewers. Kebabs are usually made of sliced lean meat from veal, mutton or chicken or mixes of them. The lean meat has been marinated (mixture of salt, spices and oil) and the marinated meat pieces are arranged around a skewer bar. The usual quantity of meat on the skewer is 3-4 kg.

For preparing the product for consumption, the skewer is slowly rotated in a vertical position close to a source of heat. Traditionally glowing charcoal was positioned on the backside of the skewer in a metal basket. Nowadays gas elements, electro coils or infrared devices are used. The outside layers of the meat bulk, once they are sufficiently heated (slightly crispy), are carefully trimmed off as thin slices. In doing so, the deeper layers, which are still uncooked, will be exposed to the heat and trimmed off when cooked. The process is repeated until all meat has been trimmed off. A special kebab is produced using minced or finely comminuted meat mixes similar to patty mixes. This type of kebab must be heat treated (coagulated) prior to final roasting to make sure that the big chunk of meat firmly sticks to the vertical skewer and maintains its shape and position.

Other varieties of kebabs are prepared in individual portions in fast-food outlets. These kebab types usually consist of fresh or marinated small meat dices or flakes on a skewer. Some variations can contain visible portions of vegetables (bell pepper, onions, etc) or even liver/kidney pieces. A typical marinated meat-only variety is the Greek souflaki containing veal or lamb meat which is marinated with lemon juice, herbs and garlic. Souflaki is grilled over charcoal. Another variety, where often vegetables and liver/kidney pieces are included, is known as shashlik. This type is briefly fried (browned) in little oil and simmered in a heavy sauce. These individually portioned kebab varieties are nowadays also available raw (fresh or frozen) as convenience products and prepared by customers at home.

Fresh sausages

Fresh sausages probably represent the oldest form of processed meat products. Their production could be carried out everywhere where animals were slaughtered, which produced both the meat and the casings. In the simplest way of manufacture, no tools other than knives are needed. Fresh meat and fat are mixed with salt and spices and stuffed into natural casings derived from small intestines of slaughter animals. Higher quality fresh sausages are primarily composed of lean meat and fat. In some low-cost formulations non-meat extenders are also used.

Fresh sausages products are well suited for small-scale meat processing outlets, as all ingredients including casings can be generated or procured locally. The manufacture can take place with basic meat processing tools and machinery. These sausages do not undergo heat treatment at processor level, but are roasted, fried, boiled or otherwise heat treated before consumption upon demand by consumers or by consumers themselves.

Cured Meat Cuts

Cured meat cuts are made of entire pieces of muscle meat and can be sub-divided into two groups, cured-raw meats and cured-cooked meats. The curing for both groups, cured-raw and cured-cooked, is in principle similar: The meat pieces are treated with small amounts of nitrite, either as dry salt or as salt solution in water.

The difference between the two groups of cured meats is:

- Cured-raw meats do not undergo any heat treatment during their manufacture. They undergo a processing period, which comprises curing, fermentation and ripening in controlled climatized conditions, which makes the products palatable.
- Cured-cooked meats, after the curing process of the raw muscle meat, always undergo heat treatment to achieve the desired palatability.

Raw-cooked Meat Products

The product components *muscle meat, fat* and *non-meat ingredients* which are processed raw, i.e. uncooked by comminuting and mixing. The resulting viscous mix/batter is portioned in sausages or otherwise and thereafter submitted to heat treatment, i.e. "cooked". The heat treatment induces protein coagulation which results in a typical firm-elastic texture for raw-cooked products. In addition to the typical texture the desired palatability and a certain degree of bacterial stability is achieved

Precooked-cooked Meat Products

Precooked-cooked meat products contain mixes of lower-grade muscle trimmings, fatty tissues, head meat, animal feet, animal skin, blood, liver and other edible slaughter by-products. There are two heat treatment procedures involved in the manufacture of precooked-cooked products. The *first heat treatment* is the precooking of raw meat materials and the *second heat treatment* thecooking of the finished product mix at the end of the processing stage. Precooked-cooked meat products are distinguished from the other categories of processed meat products by precooking the raw materials prior to grinding or chopping, but also by utilizing the greatest variety of meat, animal by-product and non-meat ingredients

Raw-fermented Sausages

Raw-fermented sausages are uncooked meat products and consist of more or less coarse mixtures of lean meats and fatty tissues combined with salts, nitrite (curing agent), sugars and spices and other non-meat ingredients filled into casings. They receive their characteristic properties (flavour, firm texture, red curing colour) through fermentation processes. Shorter or longerripening phases combined with moisture reduction ("drying") are necessary to build-up the typical flavour and texture of the final product. The products are not subjected to any heat treatment during processing and are in most cases distributed and consumed raw.

Dried Meat Products

Dried meat products are the result of the simple *dehydration* or *drying of lean meat* in natural conditions or in an artificially created environment. Their processing is based on the experience that dehydrated meat, from which a substantial part of the natural tissue fluid was evaporated, will not easily spoil. Pieces of lean meat without adherent fat are cut to a specific uniform shape that permits the gradual and equal drying of whole batches of meat. Dried meat is not comparable to fresh meat in terms of shape and sensory and processing properties, but has significantly longer shelf-life. Many of the nutritional properties of meat, in particular the protein content, remain unchanged through drying.

Meat Processing Hygiene

Meat processing hygiene is part of Quality Management (QM) of meat plants and refers to the hygienic measures to be taken during the various processing steps in the manufacture of meat products. Regulatory authorities usually provide the compulsory national framework for food/meat hygiene programmes through laws and regulations and monitor the implementation of such laws. At the meat industry level, it is the primary responsibility of individual enterprises to develop and apply efficient meat hygiene programmes specifically adapted to their relevant range of production.

Operations in meat processing plants comprise the manufacture of value-added meat products from primary products of meat origin and non-meat origin. There are three principles of meat hygiene, which are crucial for meat processing operations.

- Prevent microbial contamination of raw materials, intermediate (semi-manufactured) goods and final products during meat product manufacture through absolute cleanliness of tools, working tables, machines as well as hands and outfits of personnel.
- Minimize microbial growth in raw materials, semi-manufactured goods and final products by storing them at a low temperature.

— Reduce or eliminate microbial contamination by applying heat treatment at the final processing stage for extension of shelf life of products (except dried and fermented final products, which are shelf-stable through low a_w and pH)

The above three principles guide meat hygiene programmes in the further processing of meat. However, meat processing hygiene is more complex. In particular, the hygienic treatment of meat before reaching the processing stage is of utmost importance for the processing quality of the meat. Failures in slaughter hygiene, meat cutting and meat handling/transportation and in the hygiene of by-products and additives will all contribute to quality losses and deterioration of the final processed meat products.

Highly contaminated raw meat is unsuitable for further processing. Final products made from hygienically deficient raw meat materials are unattractive in colour, tasteless or untypical in taste with reduced shelf life due to heavy microbial loads. Moreover, there is also the risk of presence of food poisoning microorganisms, which can pose a considerable public health hazard.

In the light of growing consumer consciousness as well as regionalization and globalization in trade, quality conscious meat plants need internal quality control/quality management schemes not only for the final products but also for the raw materials and the various processing steps.

Such Quality Management Schemes (QM) have technical and hygienic components. Technical aspects encompass product composition, processing technologies, packaging, storage and distribution.

References

FAO. 1990. Manual of simple methods of meat preservation. *FAO Animal Production and Health Paper* No. 79. Rome, FAO.

Hinman, Robert B., Harris, Robert B. *The Story of Meat.* Swift & Company, 1939. Katherine.

Karmas, E. 1976. *Processed meat technology*. New Jersey, USA, Noyes Data Corporation.

Winrock International. 1992. *Animal agriculture in developing countries: technology dimensions*. Morrilton, AR, USA, Winrock International.

7

Heat Treatment of Food

The processing of foods by heat (or heat treatment) is the most important conservation technique of long duration. It aims to destroy or partially or totally inhibit enzymes and microorganisms, whose presence or proliferation could alter the food in question or make it unfit for human consumption.

The effect of heat treatment is related to the couple time / temperature. In general, the higher the temperature is high and the longer the term, the greater the effect will be important. However, we must also take into account the thermal resistance of microorganisms and enzymes and is highly variable.According to the objective, there are several techniques of food preservation by heat treatment such as sterilization, pasteurization, thermisation, cooking and laundering.

It is important to recognise that there are various degrees of preservation by heating, and that commercial heat-preserved foods are not truly sterile. A few terms must be defined and understood.

— Sterilisation refers to the complete destruction of microorganisms. Because of the resistance of certain bacterial spores to heat, this frequently requires a treatment of at least 121°C of wet heat for 15 min or its equivalent and that every particle of the food must receive this heat treatment. For example if a can of food is to be sterilised, then immersing it into a 121°C pressure cooker or retort for 15 min will not be sufficient because of the relatively slow rate of heat transfer through the food into the can. Depending on the size of the can, the effective time to achieve true sterility may be several hours. During this time there can be many changes in the food which reduce its quality. Fortunately, many foods need not be completely sterile to be safe and have keeping quality.

— Commercially Sterile: The term commercially sterile or the word "sterile" means that degree of sterilisation at which all pathogenic and toxin forming organisms have been destroyed, as well as all other types of organisms which if present could grow in the product and produce spoilage under normal handling and storage conditions.

Commercially sterilised foods may contain a small number of beat resistant bacterial spores, but these will not normally multiply in the food supply. However, if they were isolated from the food and given special environmental conditions, they could be shown to be alive. This term describes the condition that exists in most of canned or bottled products manufactured under Good Manufacturing Practices procedures and methods; these products generally have a shelf-life of two years or more. Even after longer periods, so-called deterioration is generally due to texture or flavour changes rather than to microorganism growth.

Heat Treatments

Heat sufficient to destroy microorganisms and food enzymes also generally affects other properties of foods adversely. The mildest heat treatments that guarantee freedom from pathogens and toxins and produce the desired storage life will be the heat, treatments of choice. To select a safe heat-preservation treatment, the following facts are taken into consideration.

— Time-temperature combination required to inactivate the most heat-resistant pathogens and spoilage organisms in a particular food

— Heat-penetration characteristics in a particular food, including the can or container of choice if it is packaged.

  — Processors must provide the heat treatment which will ensure that the remotest particle of food in a batch or within a container will receive sufficient heat, for a sufficient time, to inactivate both the most resistant pathogen and the most resistant spoilage organisms if they are to achieve sterility or commercial sterility, and to inactivate the most heat-resistant pathogen if pasteurisation for public health purposes is the goal.

— Different foods will support growth of different pathogens and different spoilage organisms, and so the targets will vary depending on the food to be heated.

Food acidity/pH value has a tremendous impact on the target in heat preservation/ processing. Sequence of operations employed in heat preservation of foods (fruit and vegetables, etc.)

Even after the time and temperature required to destroy target organisms are known from thermal death curves and a sufficient margin of safety has been calculated, a problem remains how to ensure that every particle of food receives the required heat treatment. This becomes a problem of heat transfer, that is, heat penetration into and throughout the can or mass of food. If cans are heated from the outside, as would be the case if they are submerged within a retort, the larger the can, the longer it will take to heat the center portion of the can to any desired temperature. There are several other factors besides the size and shape of a can that affect heat penetration into the food within it. Principal among these factors is the nature and consistency of the food itself. This determines, for example, whether heat will reach the center by straight conduction or will be speeded by some convection within a can.

Heat energy is transferred by conduction, convection, and radiation. In retorts used in canning, conduction and convection are important. Conduction is the method of heating in which the heat moves from one particle to another by contact, in more or less straight lines. In the case of conduction, the food does not move in the can and there is no circulation to stir hot food with cold food. Convection, on the other hand, involves movement in the mass being heated. In natural convection the heated portion of the food becomes lighter in density and rises; this sets up circulation within the can. This circulation speeds temperature rise of the entire contents of the can. Forced convection occurs when circulation is promoted mechanically.

A liquid food such as canned tomato juice can be readily set into convection heating motion in addition to the heating by conduction it receives through the can wall. On the other hand, a solid food such as corned beef hash is too viscous to circulate, and so it will be virtually completely heated by conduction through the can wall and through itself. A product containing free liquid and solid, such as a can of pears within a sugar syrup, will be intermediate and will rise in temperature from a combination of conduction and convection; conduction through the fruit and convection from the moving syrup. Convection heating is far more rapid than conduction heating and so, other things being equal, if cans of these three products were placed in the same retort, uniform complete heating would be expected to be reached first in the tomato juice, second in the canned pears, and last in the corned beef hash.

When heat is applied from the outside, as in retorting, the food nearest the can surfaces will reach sterilisation temperature sooner than the food nearer the center of the can. The point in a can or mass of food which is last to reach the final heating temperature is designated the "cold point" within the can or mass. In a can of solid food heated by conduction the cold point is located in the very center of the can. However, in foods that undergo convection heating, unless the cans are agitated, the cold point is somewhat below the dead center of the can.

To ensure that commercial sterilisation is achieved, sufficient time must be allowed for the cold point of cans to reach the sterilisation temperature and remain there for the required time interval to destroy the most resistant bacterial spores. If 12 D values are indicated and this corresponds, for example, to 121°C for 2.5 min in a particular system, and if we ensure that the cold point of cans receives 121°C for 2.5 min, or an equivalent heat treatment, we are assured that every other region within the can has been adequately heated.

Protective Effect on Microorganisms

Several constituents of foods protect microorganisms to various degrees against heat. For example, sugar in high concentration protects bacterial spores, and canned fruit in a sugar syrup generally requires a higher temperature or longer time for sterilisation than the same fruit without sugar. Starch and protein in foods generally act somewhat like sugar. Fats and oils have a great protective effect on microorganisms and their spores by interfering with the penetration of wet heat. As has been noted, wet heat at a given temperature is more lethal than dry heat, because moisture is an effective conductor of heat and penetrates into microbial cells and spores.

If microorganisms are trapped within fat globules, then moisture can less readily penetrate into the cells and heating becomes more like dry heat. In the same can or food mass, organisms in the liquid phase may be quickly killed while more heating time is required for inactivation of the oil-phase flora. This makes sterilisation of meat products and fish packed in oil very difficult; the severe heat treatments required often adversely affect other food constituents. Likewise, because there is more fat and more sugar in ice cream mix than there is in milk, ice cream mix must be pasteurised at a higher temperature or for a longer time than milk to accomplish equivalent bacterial destruction.

In addition to any direct protective effects food constituents may have on microorganisms, there are indirect effects related to differences in heat conductivity rates through different food materials. Fat, for example, is a poorer conductor of heat than is water. Further, and often more important, are the effects related to food consistency and its influence on whether conduction or convection heating will take place. If sufficient starch or other thickener is added to a food composition to convert it from a convection heating system to a conduction heating system, then in addition to any direct microbial protection, there will be a slowing down of the heat penetration rate to the cold point within the container or food mass, and this will protect microorganisms.

Because common starches in solution thicken upon heating, foods supplemented with starch have a reduced rate of convection within cans during retorting and require longer retort times. Special starches have been developed which do not thicken on early heating but instead thicken on later heating or on cooling. Foods supplemented with these starches retain maximum convection rates in the retort, permitting shortened retort times and less heat damage. Then upon cooling, the starch imparts the desired thickening. In a typical application, a product like chow mein can be heated with less softening of the vegetables from excessive heating yet possess the desired viscosity in the liquid phase. The size and type of container that holds the food during thermal processing can also affect the sterilisation process.

Thin flexible pouches allow faster heat penetration into the cold center when compared to a cylindrical shape of a can, for example. This means that less heat is required for an equivalent lethality in pouches. Often higher quality product can be achieved from pouches than cans but at a higher container cost. The heat transfer properties of the container can also affect processing time. Thus, metal cans transfer heat more readily than plastic cans, resulting in shorter processing times.

Different Temperature-time

Different temperature-time combinations that are equally effective in microbial destruction can differ greatly in their damaging effect on foods. This is of the greatest practical importance in modern heat processing and is the basis for several of the more advanced heat preservation methods. If the time-temperature combinations required for destruction of C. botulinum in low-acid media are taken from thermal death curves, the following will be found to be equally effective:

— 0.78 min at 127°C,

— 10 min at 116°C,

— 1.45 min at 124°C,

— 36 min at 110°C,

— 2.78 min at 121°C,

— 150 min at 104°C,

— 5.27 min at 118°C,

— 330 min at 100°C.

This illustrates the simple relationship that the higher the temperature the less time is required for microbial destruction. This principle holds true for all types of microorganisms and spores. On the other hand, foods are not equally resistant to these combinations, and the more important factor in damaging the colour, flavour, texture, and nutritional value of foods is long time rather than high temperature. If we were to inoculate milk with C. botulinum and then heat samples for 330 min at 100°C, 10 min at116°C, and less than 1 min at 12M, equal microbial destruction would occur in all three samples, but heat damage to the milk would be enormously different.

The sample heated for 330 min would be thoroughly cooked in flavour and brown in colour. The 10-min sample would be almost as bad. The 1-min sample, although still somewhat overheated, would not be far different from unheated milk. This difference in sensitivity to time and temperature between microorganisms and various foods is a general phenomenon. It applies to milk, meat, juices, and generally all other heat-sensitive food materials. The greater relative sensitivity of microorganisms than food constituents to high temperatures can be quantitatively defined in terms of different temperature coefficients for their destruction.

Thus, whereas each increase of 10°C in temperature approximately doubles the rate of chemical reactions contributing to food deterioration, each 10°C increase, above the maximum temperature for growth, produces approximately a tenfold increase in the rate of microbial destruction. Since higher temperatures permit use of shorter times for microbial destruction, and shorter times favour food quality retention; high-temperature-short-time heating treatments rather than low temperature-long-time heating treatments are used for heatsensitive foods whenever possible. In pasteurising certain acid juices, for example, the industry formerly used treatments of about 63°C for 30 min.

Today, flash pasteurisation at 100°C for 1 min, 100°C for 12 sec, or 121°C for 2 sec is the common practice. Although bacterial destruction is very nearly equivalent, the 12 PC 2-sec treatments give the best quality juice with respect to flavour and vitamin retention. Such short holding times, however, require special equipment which is more difficult to design and generally is more expensive than that needed for processing at 63°C.

Thermal Death Curves

Bacteria are killed by heat at a rate that is very nearly proportional to the number present in the system being heated. This is referred to as a logarithmic order of death, which means that under constant thermal conditions the same percentage of the bacterial population will be destroyed in a given time interval, regardless of the size of the surviving population. In other words, if a given temperature kills 90% of the population in the first minute of heating, 90% of the remaining population will be killed in the second minute, 90% of what is left will be killed in the third minute, and so on. The logarithmic order of death also applies to bacterial spores, but the slope of the death curve will differ from that of vegetative cells, reflecting the greater heat resistance of spores.

The concept of the "D value," which is defined as the time in minutes at a specified temperature required to destroy 90% of the organisms in a population. Thus, the D value, or decimal reduction time, decreases the surviving population by one log cycle. For example if a quantity of food in a can contained one million organisms and it received heat for a time equal to four D values, then it would still contain 100 surviving organisms. If there were 100 such cans in a retort initially and the retort provided heat for a period equivalent to 7D values, then it would be expected that the 100 cans with a total initial bacterial population of 100 million organisms would still contain 10 surviving organisms. Statistically, these 10 organisms should be distributed among the cans.

Obviously, no container can have a fraction of an organism although the 100 cans will average 0.1 organisms per can. In this case, 10 of the cans probably will have one organism each and could possibly ultimately spoil, while 90 of the cans would be sterile.

A thermal death time curve for a specific organism in a specific medium or food provides data on the destruction times for a defined population of that organism at different temperatures. Two terms to characterise thermal death time curves. These are the "z value" and the "F value." The z value is the number of degrees required for a specific thermal death time curve to pass through one log cycle (change by a factor of 10). It is also the negative slope index of the thermal death time curve.

Different organisms in a given food will have different z values, which characterise resistance of the population to changing temperature. Similarly, a given organism will have different z values in different foods. The F value is defined as the number of minutes at a specific temperature required to destroy a specified number of organisms having a specific z value. Thus, the F value is a measure of the capacity of a heat treatment to sterilise. Since F values represent the number of minutes to diminish a population with a specific z value at a specific temperature, and z values as well as temperatures vary, it is convenient to designate a reference F value.

Such a reference is the F0 value, which equals the number of minutes at 121°C (250°F) required to destroy a specified number of organisms whose z value is 10°C (18°F). If such a population is destroyed in 6 min at 121°C, then the heat treatment was equal to an F0 of 6. Other temperatures for different times can have the same lethality as this heat treatment. If they do, then they also can be described as having an F0 value of 6. If they have less lethality, they have an F0 value of less than 6 and vice versa.

The F0 value of a heat treatment is thus a measure of its lethality, and the F0, value is also known as the "sterilisation value" of the heat treatment. F0 is a common term in the canning industry and other areas utilising heat processes. Not only do different amounts of heating provide different F0 values but the F0 requirements of various foods differ and are a measure of the ease or difficulty with which these foods can be heat-sterilised. Thermal death time curves have been carefully determined for many important pathogens and food spoilage organisms.

Two such curves for Putrefactive Anaerobe 3679 and Bacillus steareothermophilus They tell us how long it takes to kill these organisms (under defined conditions) at a chosen Temperature. Thus, for example, it would take about 60 min at 104°C (220°F) to kill a specified number of spores of PA 3679. On the other hand, at a temperature of 121°C (250°F) these spores are killed in little over 1 min. The conditions that must be defined to make a thermal death curve meaningful and applicable to food processing are many. The requirement for a greater heat treatment the larger the initial microbial population is inherent in the logarithmic order by which bacteria die.

Effect of Graphical Method

This method also known as the 'General method', first suggested by Bigelow and his co-workers for process calculation, is essentially a graphical procedure for integrating the lethal effects of various time and temperature relationships existing at the end point during heating and cooling. Each temperature represented by a point on the heat penetration curve is considered to have a sterilising or lethal value.

The method is briefly as follows:

— The thermal death time (TDT) for the most resistant spoilage organism likely to be encountered is determined in the food being canned. Thermal death times from this curve are converted into lethal rates for the various heating temperatures. The lethal rate is the reciprocal of the thermal death time (l/TDT).

— The heat penetration curve for the food involved is determined.

The product of lethal rate and time is equal to lethality, the area beneath the curve may be expressed directly in units of lethality. A unit sterilisation area is defined as the area on a lethality curve, which just represents complete sterilisation. To determine what process time must e employed to give unit lethality, the cooling position of any given lethality curve may be shifted to the right or to the left so as to give an are equal to 1. An area equal to 20 small squares (0.01 x.5 x 20 = 1) is equal to unit sterilisation area. The area under the lethality curve may be measured either by counting squares or by using a planimeter.

The area OCBADEFO representing 47 minutes process, is 38.4 squares, which is 1.92 times of the unit sterilisation area. This means that the process is considerably longer than what is necessary. To find that time required for unit sterilisation, cooling curves similar to the original cooling curve, are drawn at several points along the heating curve. The areas OCFO and OCBEFO for starting at 25 and 35 minutes are 8.4 and 1.94 squares, which correspond to 42 % and 97 % of unit sterilisation area. By such graphical interpolation method, it will be found that a process of 35.5 mil lutes will result in complete destruction. This is a trial and error procedure.

Pasteurisation Process

The heating of every particle of milk or milk product to a specific temperature for a specified period of time without allowing recontamination of that milk or milk product during the heat treatment process. There are two distinct purposes for the process of milk pasteurisation:

— *Public Health Aspect:* to make milk and milk products safe for human consumption by destroying all bacteria that may be harmful to health (pathogens)

— *Keeping Quality Aspect:* to improve the keeping quality of milk and milk products. Pasteurisation can destroy some undesirable enzymes and many spoilage bacteria. Shelf life can be 7, 10, 14 or up to 16 days.

Technological Principles

Physical and chemical factors which influence pasteurisation process are the following:

— temperature and time;

— acidity of the products;

— air remaining in containers.

Pasteurisation processes. In pasteurising certain acid juices for example, there are two categories of processes:

— Low pasteurisation where pasteurisation time is in the order of minutes and related to the temperature used; two typical temperature/time combinations are as following: 63° C to 65° C over 30 minutes or 75° C over 8 to 10 minutes.

Pasteurisation temperature and time will vary according to:

— nature of product; initial degree of contamination;

— pasteurised product storage conditions and shelf life required.

In this first category of pasteurisation processes it is possible define three phases:

— heating to a fixed temperature;

— maintaining this temperature over the established time period (= pasteurisation time);

— cooling the pasteurised products: natural (slow) or forced cooling.

— Rapid, high or flash pasteurisation is characterised by a pasteurisation time in the order of seconds and temperatures of about 85° to 90° C or more, depending on holding time. Typical temperature/time combinations are as follows:

— 88° C (190° F) for 1 minute;

— 100° C for 12 seconds;

— 121°C for 2 seconds.

While bacterial destruction is very nearly equivalent in low and in high pasteurisation processes, the 121° C/2 seconds treatment give the best quality products in respect of flavour and vitamin retention. Such short holding times, however, require special equipment which is more difficult to design and generally is more expensive than the 63-65 ° C/30 minutes type of processing equipment.

In flash pasteurisation the product is heated up rapidly to pasteurisation temperature, maintained at this temperature for the required time, then rapidly cooled down to the temperature for filling, which will be performed in aseptic conditions in sterile receptacles. Taking into account the short time and rapid performance of this operation, flash pasteur-isation can only be achieved in continuous process, using heat exchangers. Industrial applications of pasteu-risation process are mainly used as a means of preservation for fruits and vegetable juices and specially for tomato juice.

Thermopenetration Problem

The thermopenetration problem is extremely important, especially in the case of the pasteurisation of products packed in glass containers because it is the determining factor for the success of the whole operation. During pasteurisation it is necessary that a sufficient heat quantity is transferred through the receptacle walls; this is in order that the product temperature rises sufficiently to be lethal to microorganisms throughout the product mass.

The most suitable and practical method to speed up thermopenetration is the movement of receptacles during the pasteurisation process. Rapid rotation of receptacles around their axis is an efficient means to accelerate heat transfer, because this has the effect, among others of rapidly mixing the contents. The critical speed of for this movement is generally about 70 rotations per minute (RPM). This enables a more uniform heating of products, reducing heating time and organoleptic degradation.

Pasteurisation Methods

There are two basic methods, batch or continuous.

Batch Pasteurisation

One of the earliest and simplest methods of effectively pasteurising liquid foods, such as milk, is to heat the food in a vat with mild agitation. Raw milk commonly is pumped into a steam-heated jacketed vat, brought to temperature, held for the prescribed time, and then pumped over a platetype cooler prior to bottling or cartoning. Milk must be quickly brought to 62.8°C (145°F), held at this temperature for 30 min, and rapidly cooled. In addition to destroying common pathogens, this heat treatment also inactivates the enzyme lipase, which otherwise would quickly cause the milk to become rancid. Batch pasteurisation, also known as the holding method of pasteurisation, is still widely practiced in some parts of the world, but it has largely given way to hightemperature-short-time continuous pasteurisation.

Continuous Method

Continuous process method has several advantages over vat method, the most important being time and energy saving. For most continuous processing, a high temperature short time (HTST) pasteuriser is used. The heat treatment accomplished using a plate heat exchanger. This piece of equipment consists of a stack of corrugated stainless steel plates clamped together in a frame. There are several flow patterns that can be used. Gaskets are used to define the boundaries of the channels and to prevent leakage. The heating medium can be vacuum steam or hot water.

High-Temperature-Short-Time Pasteurisation High-temperature-short-time (HTST) pasteurisation of raw milk employs a temperature of at least

71.7°C (161°F) for at least 15 see. This is equivalent in bacterial destruction to the batch method. In HTST pasteurisation raw milk held in a cool storage tank is pumped through a plate-type heat exchanger and brought to temperature. The key to the process rests in ensuring that every particle of the milk remains at not lower than 71.M for no less than 15 see.

This is accomplished by pumping the heated milk through a holding tube of such length and diameter that it takes every milk particle at least 15 see to pass through the tube. At the end of the tube is an accurate temperature-sensing device and valve. Should any milk reach the end of the holding tube and be down in temperature even one degree, a flow diversion valve checks this flow of milk and sends it back through the heat exchanger once again to be reheated. In this way no milk escapes the required heat treatment. Frequent checks of the equipment are made by authorised milk inspectors to help ensure its proper operation. After emerging from the holding tube, the milk is cooled and may be cartoned or bottled.

Cooling not only prevents further heat damage to the milk but also retards subsequent bacterial multiplication since the milk is not sterile. Pasteurisation by the HTST method is not limited to milk and is widely used in the food industry. However, times and temperatures vary in accordance with the effects of different foods on microorganism survival and the heat sensitivities of these foods.

Equipment of HTST Pasteuriser

Balance Tank

The balance, or constant level tank provides a constant supply of milk. It is equipped with a float valve assembly which controls the liquid level nearly constant ensuring uniform head pressure on the product leaving the tank. The overflow level must always be below the level of lowest milk passage in regenerator. It, therefore, helps to maintain a higher pressure on the pasteurised side of the heat exchanger. The balance tank also prevents air from entering the pasteuriser by placing the top of the outlet pipe lower than the lowest point in the tank and creating downward slopes of at least 2%. The balance tank provides a means for recirculation of diverted or pasteurised milk.

Regenerator

Heating and cooling energy can be saved by using a regenerator which utilises the heat content of the pasteurised milk to warm the incoming cold milk. Its efficiency may be calculated as follows: % regeneration = temp. increase due to regenerator/total temp. increase For example: Cold milk entering system at 4° C, after regeneration at 65° C, and final temperature of 72° C would have an 89.7% regeneration:

$$\frac{65-4}{72-4} = 89.7$$

Holding Tube

Must slope upwards ¼"/ft. in direction of flow to eliminate air entrapment so nothing flows faster at air pocket restrictions.

Indicating Thermometer

The indicating thermometer is considered the most accurate temperature measurement. It is the official temperature to which the safety thermal limit recorder (STLR) is adjusted. The probe should sit as close as possible to STLR probe and be located not greater than 18 inches upstream of the flow diversion device.

Flow Diversion Device (FDD)

Also called the flow diversion valve (FDV), it is located at the downstream end of the upward sloping holding tube. It is essentially a 3-way valve, which, at temperatures greater than 72° C, opens to forward flow. This step requires power. At temperatures less than 72° C, the valve recloses to the normal position and diverts the milk back to the balance tank.

It is important to note that the FDD operates on the measured temperature, not time, at the end of the holding period. There are two types of FDD: single stem—an older valve system that has the disadvantage that it can't be cleaned in place dual stem—consists of 2 valves in series for additional fail safe systems. This FDD can be cleaned in place and is more suited for automation. The hatched area in the graph above represents the area under the time temperature curve that is taken into consideration for thermal lethality calculations

Vacuum Breaker

At the pasteurised product discharge is a vacuum breaker which breaks to atmospheric pressure. It must be located greater than 12 inches above the highest point of raw product in system. It ensures that nothing downstream is creating suction on the pasteurised side.

Use of Auxiliary Equipment

Booster Pump

It is centrifugal "stuffing" pump which supplies raw milk to the raw regenerator for the balance tank. It must be used in conjunction with pressure differential controlling device and shall operate only when timing pump is operating, proper pressures are achieved in regenerator, and system is in forward flow.

Homogeniser

The homogeniser may be used as timing pump. It is a positive pressure pump; if not, then it cannot supplement flow. Free circulation from outlet to inlet is required and the speed of the homogeniser must be greater than the rate of flow of the timing pump.

Holding Time

When fluids move through a pipe, either of two distinct types of flow can be observed. The first is known as turbulent flow which occurs at high velocity and in which eddies are present moving in all directions and at all angles to the normal line of flow. The second type is streamline, or laminar flow which occurs at low velocities and shows no eddy currents. The Reynolds number, is used to predict whether laminar or turbulent flow will exist in a pipe:

— Re < 2100 laminar

— Re > 4000 fully developed turbulent flow

There is an impact of these flow patterns on holding time calculations and the assessment of proper holding tube lengths. The holding time is determined by timing the interval for an added trace substance (salt) to pass through the holder. The time interval of the fastest particle of milk is desired. Thus the results found with water are converted to the milk flow time by formulation since a pump may not deliver the same amount of milk as it does water.

Ultra High Temperature Pasteurisation

The third is UHT, ultra- high temperature pasteurisation. This method is used mainly for coffee creamers and boxed juices with the exception of

Europe. They pasteurise milk in this way. These products are brought to over the boiling point, 250F (under pressure) for only a fraction of a second. After this is done, there is no need to refrigerate, because it sterilises the product. Sometimes the products can have a "cooked" taste that can be detected after being brought to such a high temperature.

Advantages and Disadvantages

The advantages to be obtained from the efficient pasteurisation of milk may be summarise as follows:

— Milk is rendered 'safe' as all pathogens are destroyed: It would not be possible to obtain safe milk supply for the community at all times without using this process. This is the simplest way by which safety may be guaranteed.

— The bulk of the non-pathogenic bacteria are also destroyed: It is probably quite true to state that approximately only I % of non-pathogenic bacteria rcm-ain after milk treatment by any officially approved method of pasteurisation while, with some methods, a greater percentage is destroyed. The percentage which remains may be greatly increased when conditions are such as will allow large numbers of heatloving and heat-resistant organisms to be present. Those microorganisms which do survive are usually sporefo-rmers. These do not affect the keeping quality of the liquid or the health of the consumer.

— The useful life of the milk is prolonged: This is important to the distributor who, in many cases, has to rely upon a supply of raw milk which has been transp-orted to long distances before treatment. In addition to this, the consumer might be unable to obtain milk if heat treatment was forbidden.

There are certain alleged disadvantages. The voicing of these seems to depend upon the outlook or the opinion of the person concerned. These are stated to be:

— Pasteurisation cannot render dirty milk clean: Although some milk producers may not think so, clean milk is a prime necessity for efficient pasteurisation. Dirty, raw milk is likely to contain heat-loving or heat-resistant microorganisms in abundance, apart from other non-pathogenic bacteria. Nonpathogenic bacteria will reduce the keeping quality of the pasteurised product.

— Pasteurisation reduces the cream-line: A temperature of 145°F. (62.7°C.) for 30 minutes injures the cream-line and a 10% to 14% reduction may be expected if the low temperature process is employed. With the high temperature short time process, the reduction can vary from 10% to 20%. The cream-line may be preserved to some extent by the rapid cooling of the hot milk. The critical temperatures are between 110°F. (43.3°C.) and 60°F. (15.5°C.). It is essential to cool the milk quickly between these limits if the cream-line is to be conserved. Generally, the lower the temperature to which milk is cooled, the better will be the cream-line. Chemically, however, the cream content of the milk remains unaltered.

— Chemical composition of the milk is altered in accordance with the temperatures to which it is heated. Generally, with an approved pasteurisation process, the position is as follows:

— 5% to 10% of the albumin is precipitated.

— Usually about 5% of the calcium and phosphates contained in the milk are precipitated in an insoluble form.

— Proteins and enzymes begin to be affected at 150°F. (65-5°C.). They are completely destroyed at very high temperatures for short periods.

— The lactose is unaffected by the process.

  — The vitamin content of the milk is altered to a varying degree as follows:

  — Vitamin A - not affected by pasteurisation.

  — Vitamin B1- not heat tolerant, a reduction of approximately 10% to 25% occurring.

  — Vitamin B-Complex - not affected.

  — Vitamin C - heat sensitive and destroyed to a greater or lesser degree by the process according to the method used. With the low-temperature process, a reduction of upto 25% may be expected, but the short-time process is less harmful in this respect, not more than 5% being destroyed during the treatment. It would appear that the amount destroyed depends upon previous exposure of the milk to light, particularly sunlight and the degree of oxidation which occurs, this latter process being accelerated by the catalytic actions of metals. There is, however, little action of this kind with stainless steel or the other metals used in the construction of modern plants. In any case, milk is not a good source of this vitamin and it certainly cannot provide sufficient quantities of this vitamin for ones bodily needs even if it were unimpaired by the process. Other food-stuffs have to be consumed to supply the amount the body requires, irrespective of the type of milk consumed.

— Vitamin D - absolutely heat tolerant at pasteurising temperatures.

— Vitamin E -so far as is known, this vitamin is not affected by the process.

The flavour of pasteurised milk may be slightly different from that of raw milk due to the removal by filtration and during heating of the volatile substances derived from the cow and its surroundings. Overheating will, however, affect the flavour.

It would appear that a safe, long-keeping milk does not suffer from any really serious disadvantages as a result of pasteurisation. Apart from a slight reduction in the amount of vitamin C, the nutritive value is not materially impaired.

Aseptic Packaging

Combining the best attributes of paper, plastic, and aluminium, the multi-layer, high-performance aseptic package (commonly known as the "drink box") locks out light and air, seals in nutrients and flavour, and allows its contents to remain unrefrigerated for months.. The aseptic process, which goes hand-in-hand with the packaging, is a major advance over traditional canning techniques; An excellent package becomes a great package when you consider perhaps its most important but least understood benefit—the aseptic package is nothing short of an environmental hero.

The drink box is environmentally smart from the start. It is one of the best examples of minimal packaging (less material from the start) available today. And it uses far less energy to manufacture, fill, ship, and store than virtually any other comparable package on the market. the aseptic package is a lightweight, energy efficient example of minimal packaging (typically 96% product to 4% packaging, by weight) that is being recycled in a growing number of communities across the country. In addition, aseptic packaging retains flavour and nutrients, and allows traditionally perishable products (like real milk) to stay unrefrigerated and shelf-stable for up to six months.

Aseptic Processing and Packaging

Aseptic processing sterilises the food product by destroying the harmful bacteria and pathogenic microorganisms through a tightly controlled thermal process. The Enhance™ Flexible Aseptic Packaging Systems provide a sterile environment for the product to be introduced into the packaging material. The packaging film barrier properties protect the sterile product from outside contamination. This results in a packaged product that is shelf stable and does not require refrigeration. Because Aseptic processing provide more precise temperature treatment to the product, the results is a more repeatable flavour profile Aseptic Canning- It is possible to shorten sterilising times down to seconds and even fractions of a second, and for many products this results in a marked improvement in quality.

This can be done with canned foods by the methods of aseptic canning. Aseptic canning refers to a technique in which food is sterilised or commercially sterilised outside of the can and then aseptically placed in previously sterilised cans which are subsequently sealed. The point behind aseptic canning is that while food in a container will required many minutes of even hours, depending upon container size, to reach sterilising temperature, food outside a container may be passed through an efficient heat exchanger and brought to sterilisation temperature almost instantaneously. It then remains to provide sterile cans and lids and to fill the cans and seal them in an aseptic environment. Food temperatures employed may be as high as 1500C (3020F) and sterilisation takes place in 1 or 2 sec yielding food products of the highest quality.

Quick heating of liquid foods may be done in a plate-type heat exchanger or a tubular scraped-surface-type heat exchanger. This latter type consists essentially of a tube within a tube. Steam flows through the space between the tubes while food flows through the inner tube. The inner tube also is provided with a rotating shaft or mutator equipped with scraper blades prevent food from burning onto the heat exchange surface. contact with the hot surface the thin layer of food may be brought to sterilisation temperature in a second or less.

If it desired to prolong residence time beyond this, then a holding tube is added as in the case of HTST pasteurisation. Such rapid sterilisation at extremely high temperatures, as 1 or 2 sec at 1500C, sometimes is referred to as ultra-high temperature (UHT) sterilisation. The sterile food must now be quickly cooled; quick cooling can be accomplished with the same types of plate or tubular scraped-surface heat exchangers, used with refrigerants instead of steam The sterile cool food now enters the aseptic canning line. This consists of a tunnel through which cans without their lids are conveyed and sterilised by superheated steam, a sterile filling zone also heated by steam where the food enters the cans, a heated sterile atmosphere. After cans are sealed they may be further cooled with sprays of water.

Advances in Aseptic Canning

Changing the type of heat exchanger and the pumps and line leading to the aseptic canning line, chunk type foods such as chow mein or chicken a laking can be aseptically canned. In this case intimate heat contact with chunk type foods to achieve rapid sterilisation can be accomplished with direct steam injection type heat exchangers. In using direct steam injection is essential that the steam not contain any impurities that may have come from the boiler. Today, there are direct steam injectors that generate steam from softened purified water to eliminate such impurities.

Steam injection heating is not restricted to use with aseptic canning. Many foods are cooked, pasteurised, or otherwise quickly heated by this method. The food manufacturer then may use the tomato paste in the production of ketchup or the apricot in bakery products. If such large volumes were to be sterilised for keeping properties in the drums, then by the time the cold point reached sterilisation temperature the product nearer the drum walls would be excessively scorched. Such items also can be quickly sterilised efficient heat exchangers and aseptically drummed.

In this case, large chambers have been developed in which the drums and lids are sterilised under superheated steam and product filled and sealed aseptically within the chamber. This technology has advanced to the point where sterile food can be aseptically filled into previously sterilised silo tanks and tank cars. Another form of aseptic packaging utilises flexible packaging materials which are sterilised, formed, filled, and sealed in a continuous operation. In some cases the disinfectant property of hydrogen peroxide is combined with heat to make lower temperatures effective in sterilising these less heat-resistant packaging materials.

The Benefits of Aseptic Process

— Food safety

— Shelf stability independent of ingredients
— Ability to process thermally sensitive products
— More robust process
— Thermally process a product independent of package size
— More precise control over process
— Flexible use of containers varying in design/materials
— Produce a product of improved sensory quality
— Product does not require refrigeration, resulting in energy savings in storage
— Improved nutrient retention

To provide shelf stability, the aseptic packaging process does not require additives or other ingredients. Aseptic Systems combine the sterile product with the sterile package in a sterile environment. The film barrier properties protect the product from outside contamination. The impact on the product is improved flavour without preservatives, lower salt content and more home-made taste.

Flexible aseptic packaging is more cost-effective. Its added benefits are:

— product safety
— less waste material
— fewer packaging materials
— more product per case to reduce shipping costs
— maximisation of storage space
— shelf stability without adding special ingredients
— elimination of refrigeration or additives
— enhanced shelf life

Aseptic packaging is a method in which food is sterilised or commercially sterilised outside of the can, usually in a continuous process, and then aseptically placed in previously sterilised containers which are subsequently sealed in an aseptic environment. The most commercially successful form of aseptic packaging utilises paper and plastic materials which are sterilised, formed, filled, and sealed in continuous operation. The package may be sterilised with heat or a combination of heat and chemicals. In some cases, the disin-fectant property of hydrogen peroxide (H,O,) is combined with heated air or with ultraviolet light to make lower temperatures effective in sterilising these less heat-resistant packaging materials.

Coffee cream and coffee whiteners, for example, are packaged in small single-service paper packets this way, as are larger volume size milk and juice products. Quick heating of liquid foods may be done in a platetype heat exchanger or in a tubular scraped-surface heat exchanger This latter type consists essentially of a tube within a tube. Steam flows through the space between the tubes while food flows through the inner tube. The inner tube also is provided with a rotating shaft or mutator equipped with scraper blades to prevent food from burning onto the heat exchange surface. In contact with the hot surface, the thin layer of food may be brought to sterilisation temperature in 1 sec or less.

Food temperatures employed may be as high as 150'C and sterilisation takes place in 1 or 2 see, yielding food products of the highest quality, and oftcn with significant energy savings. If it is desired to prolong residence time beyond this, then a holding tube is added as in the case of HTST pasteur-isation. Such rapid sterilisation at extremely high temperatures is referred to as ultrahigh-temperature (UHT) sterilisation. The sterile food must be quickly cooled to room temperature, because at these high temperatures product quality can be impaired in seconds. Quick cooling can be accomplished with the same types of plate or tubular scraped-surface heat exchangers, used with refrigerants instead of steam.

Aseptic packaging is also used with metal cans as well as large plastic and metal drums or large flexible pouches Great quantities of food materials are used as intermediates in the production of further processed foods. This frequently requires packaging of such items as tomato paste or apricot puree in large containers such as remains at not lower than 71.M for no less than 15 see. This is accomplished by pumping the heated milk through a holding tube of such length and diameter that it takes every milk particle at least 15 see to pass through the tube.

At the end of the tube is an accurate temperature-sensing device and valve. Should any milk reach the end of the holding tube and be down in temperature even one degree, a flow diversion valve checks this flow of milk and sends it back through the heat exchanger once again to be reheated. In this way no milk escapes the required heat treatment. Frequent checks of the equipment are made by authorized milk inspectors to help ensure its proper operation. After emerging from the holding tube, the milk is cooled and may be cartoned or bottled. Cooling not only prevents further heat damage to the milk but also retards subsequent bacterial multiplication since the milk is not sterile.

Aseptic packaging facility specialises in Tetra Pak slim, square, wedge and Combi Bloc packages, with a variety of opening types. The extended shelf life and logistical benefits of aseptic packaging make it one of the fastest growing delivery vehicles for liquid products

Hot Pack

The terms hot pack or hot fill refer to the packing of previously pasteurised or sterilised foods, while still hot, into clean but not necessarily sterile containers, under clean but not necessarily aseptic conditions. The heat of the food and some holding period before cooling the closed container is utilised to render the container commercially sterile. Hot pack, as distinguished from aseptic packaging, is most effective with acid foods since lower temperatures in the presence of acid are lethal; further, at a pH of 4.6 Clostridium botulinum will not grow or produce toxin, so this health hazard is not present.

Hot pack with lowacid foods (above pH 4.6) is not feasible unless the product is recognised as being only pasteurised and will be stored under refrigeration or unless the hot pack treatment is combined with some additional means of preservation such as a very high sugar content. This is because the residual heat of the food in the absence of appreciable acid is not sufficient to guarantee destruction of spores that may be present on container surfaces or that may enter containers during filling and sealing. Even with acid foods, very definite food temperatures and holding times in the sealed containers before they are cooled for warehouse storage must be adhered to for hot pack processing to be effective. These temperatures and times depend on the specific product's pH and other food characteristics.

In home canning, when fruit and sugar are boiled together to make jam and the hot jam is poured into jars that have been previously boiled, the principle of hot pack is being employed. Home canning instructions further call for inverting the filled jars after a short time. This is to ensure that the hot acid product contacts all surfaces of the jar lid for sterilisation. However, for home canning of meats and other low-acid foods, directions always call for pressure cooking of closed containers as is done in conventional commercial retorting.

In commercial practice, acid juices such as orange, grapefruit, grape, tomato, and various acid fruits and vegetables, such as sauerkraut, commonly are hot packed following prior pasteurisation or sterilisation. Typically, acid fruits and juices are first heated in the range of about 77-100°C for about 30-60 sec, hot filled at no lower than 63oC and often closer to 93°C, and held at this temperature for 1-3 min, including an inversion before cooling.

In the case of tomato juice, a common practice is HTST heating of juice at 121°C for 0.7 min, cooling below the boiling point but not below 63oC for hot fill can sterilisation, and can holding for 3 min, including an inversion before final cooling. Precise times and temperatures depend on the pH of the particular tomato juice batch and can be confirmed by inoculated pack studies.

Heating Treatments after Packaging Food Preservation

The foregoing principles very largely determine the design parameters for heat preservation equipment and commercial practices. The food processor will employ no less than that heat treatment which gives the necessary degree of microorganism destruction.

This is further ensured by periodic inspections from the FDA or equivalent local authorities. However, the food processor also will want to use the mildest effective heat treatment to ensure highest food quality, as well as to conserve energy.

It is convenient to separate heat preservation practices into two broad categories: one involves heating of foods in their final containers, the other employs heat prior to packaging. The latter category includes methods that are inherently less damaging to food quality, especially when the food can be readily subdivided (such as liquids) for rapid heat exchange. However, these methods then require packaging under aseptic or nearly aseptic conditions to prevent or at least minimise recontamination.

Canning is a severe heat treatment and causes changes in flavour, texture and nutrients of food. During canning time required to sterilise a food depends on:

- Heat resistance of MO or enzyme
- Retort (heating) conditions

— pH of food
— Size of container
— Physical state of food

In container:

— Heat resistance of MO
— For low pH Clostridium botulinum is the most dangerous
— Under anaerobic conditions can produce an exotoxin
— Destruction of C. botulinum is the minimum requirement Commercially a 12D process is used when C. botulinum is likely to be present.

Thermal destruction of MO takes logarithmically. “12D” process involves heat treatment that will provide a 12 log reduction of C. botulinum spores at 121oC Rate of heat penetration in container is measured using a thermocouple, at the thermal center of a container which is the geometric center and is different for conduction and convection. It is denoted by F value which is used as a basis for comparing heat sterilisation procedures. It represents the total time temperature combinations F value = time (min.)

The F value can be thought of as:

— The time needed to reduce MO numbers by a multiple of D values
— Found using: F = D (log n1 - log n2) F0

Used to describe a process that operates at 121°C

Heating Food in Containers

It can be of three kinds they are

Still Retort

Max temp of 121°C to prevent food damage near can wall Long cook time

Agitating Retort

Shorter cook time Less food damage

Hydrostatic Retort

— Continuous flow of cans
— Uses hydrostatic head to control pressure
— Is an agitating system

Now we shall study about the three retorts in detail Still Retort—One of the simplest applications of heating food in containers is sterilisation of cans in a still retort, that is, the cans remain still while they are being heated. In this type of retort, temperatures above 121°C (250°F) generally may not be used or foods cook against the can walls.

This is especially true of solid type foods which do not circulate within the cans by convection, but it also can be a problem with liquid foods. Because 121°C is the upper temperature, and there is relatively little movement in the cans, the heating time to bring the cold point to sterilising temperature is relatively long; for a small can of peas it may be 40 min.

Moist foods in cans have part of their moisture converted to steam at these temperatures and produce equivalent pressures within the cans although there is little pressure difference between the inside and outside of the can when temperatures are equal. A special case is foods canned under vacuum. Then the initial pressure within a can will be less than the pressure in the retort to an extent determined by the degree of vacuum used at time of can closure.

Control of pressure differences inside and outside of cans and other containers during and following heat treatment are of obvious importance to prevent mechanical damage to containers. Several techniques are employed to prevent such damage.

If the vacuum within cans is such that retort pressures cause can collapse, a heavier gauge of steel may be required. More commonly, pressure problems are due to greater pressures within the container than on its outside. This occurs when steam pressure is too rapidly released in closing down a batch-type retort or when heated containers are too suddenly conveyed from a continuous pressure retort to atmospheric pressure.

The problem is greater in the case of glass jars than with cans; excessive internal pressure can easily blow the lids from glass jars since these generally have a weaker seal than the lids of cans. During retorting of glass jars, provisions are made for air pressure over a layer of water to balance internal and external pressures. Partial cooling of containers before releasing them from retorts is the common way to decrease internal container pressure.

Many continuous retorts, such as the agitating type, provide semi pressurised cooling zones following the heating zone just prior to release of cans to atmospheric pressure. With the increasing use of flexible packaging materials has come the sterilisation of foods in flexible plastic pouches. Here, pressure problems can be still greater than with glass jars. Overriding air pressure must be applied when the pouches are cooled after retorting so that the steam pressure inside the pouch does not cause the pouch to burst.

Agitating Retorts

Processing time can be markedly reduced by shaking the cans during heating, especially with liquid or semi liquid foods. Not only is processing time shortened, but food quality is improved. This is accomplished with various kinds of agitating retorts, one type of which is shown Part of the wall has been cut away to show the cans resting in reels which rotate and thereby shake the contents.

Forced convection within cans also depends on the degree of can filling, since some free headspace within cans is necessary for optimum food turnover within the cans. In addition to faster heating, there is less chance for food to cook onto the can walls since the can contents are in motion. Different types of agitation are possible; for example, cans may be made to turn end over end or to spin on their long axis. Depending on the physical properties of the food, one method may be more effective.

The reduction in processing times possible with agitating retorts compared with still retorts. These substantial reductions in time with associated quality advantages would not be realised in foods that heat primarily by conduction; for such foods the simpler and generally less costly still retorts may be quite satisfactory.

Agitation in container depends upon

— Temperature of retort

— Temp. difference between product and heating medium Shape of container Type of container

— Metal vs. glass vs. plastic

Pressure Considerations

Whether still or agitating retorts are used, the high temperatures required for commercial sterilisation commonly are obtained from steam under pressure. Steam pressures of approximately 10, 15, and 20 psi (above atmospheric pressure) are required for heating at 116°Q 121°C, and 127°C (1 psi = 0.07 k g/CM^2 = 6895 pascals).

Moist foods in cans have part of their moisture converted to steam at these temperatures and produce equivalent pressures within the cans although there is little pressure difference between the inside and outside of the can when temperatures are equal.

A special case is foods canned under vacuum. Then the initial pressure within a can will be less than the pressure in the retort to an extent determined by the degree of vacuum used at time of can closure. Control of pressure differences inside and outside of cans and other containers during and following heat treatment are of obvious importance to prevent mechanical damage to containers.

Several techniques are employed to prevent such damage. If the vacuum within cans is such that retort pressures cause can collapse, a heavier gauge of steel may be required. More commonly, pressure problems are due to greater pressures within the container than on its outside.

This occurs when steam pressure is too rapidly released in closing down a batch-type retort or when heated containers are too suddenly conveyed from a continuous pressure retort to atmospheric pressure. The problem is greater in the case of glass jars than with cans; excessive internal pressure can easily blow the lids from glass jars since these generally have a weaker seal than the lids of cans. During retorting of glass jars, provisions are made for air pressure over a layer of water to balance internal and external pressures.

Partial cooling of containers before releasing them from retorts is the common way to decrease internal container pressure. Many continuous retorts, such as the agitating type, provide semi pressurised cooling zones following the heating zone just prior to release of cans to atmospheric pressure.

With the increasing use of flexible packaging materials has come the sterilisation of foods in flexible plastic pouches. Here, pressure problems can be still greater than with glass jars. Overriding air pressure must be applied when the pouches are cooled after retorting so that the steam pressure inside the pouch does not cause the pouch to burst. In addition, a uniform heat treatment requires that the pouches be evenly exposed to the heating medium rather than be allowed to contact each other and pile up.

One means of better controlling pouches during retorting is to sandwich the pouches between rigid supports Plastic pouches require shorter retort times since heat penetration through the thin pouches is quite rapid. This, in turn, can produce high quality products and save on energy costs. All plastic rigid "cans" are also being used to retort foods. These cans have the advantage that they can be reheated in a microwave oven. They require many of the same types of handling requirements as flexible pouches.

Hydrostatic Cooker and Cooler

Continuous retorts (usually of the agitating type) are pressuretight and built with special valves and locks for admitting and removing cans from the sterilising chamber. Without these, pressure conditions would not be held constant and sterilising temperatures could not be closely controlled. Another type of continuous pressure retort, which is open to the atmosphere at the inlet and outlet ends, is the hydrostatic pressure cooker and cooler. This type of heating equipment consists essentially of a "U" tube with an enlarged lower section.

Steam is admitted to the enlarged section and hot water fills one of the legs of the "U" while cool water fills the other leg. Cans are carried by a chain conveyor down the hot water leg, through the steam zone, which may involve an undulating path to increase residence time, and up the cool water leg. These legs are sufficiently high to produce a hydrostatic head pressure to balance the steam pressure in the sterilising zone. If a temperature of 127°C is used in the sterilising zone, then this would be equal to a pressure of about 140,000 pascals (20 psi) above atmospheric pressure which would be balanced by water heights of about 14 m (46 ft) in the hot and cold legs.

As cans descend the hot water leg and enter the steam zone, their internal pressure increases as food moisture begins to boil. But this is balanced by the increasing external hydrostatic pressure. Similarly, as highpressure cans pass through the water seal and ascend the cool water leg, their gradually reduced internal pressure is balanced by the decreasing hydrostatic head in this cool leg. In this way cans are not subjected to sudden changes in pressure. For this reason the system also is well suited to the retorting of foods and beverages in jars and bottles.

Where sterilising temperatures above 100°C are needed, steam under pressure generally is the heat exchange medium, and vessels capable of withstanding pressure add to the cost of equipment. Another method, employs direct flame to contact cans as the cans are rotated in the course of being conveyed past gas jets. Excellent rates of heating are achieved with high product quality and reduced costs, but commercial experience with this type of system is still somewhat limited.

Cooking

Cooking is the process of preparing food, by the analog skills, often with the use of heat. Cooking techniques and ingredients vary widely across the world, reflecting unique environmental, economic, and cultural traditions. Cooks themselves also vary widely in skill and training. Cooking can also occur through chemical reactions without the presence of heat, most notably as in Ceviche, a traditional South American dish where fish is cooked with the acids in lemon or lime juice. Sushi also utilizes a similar chemical reaction between fish and the acidic content of rice glazed with vinegar.

Preparing food with heat or fire is an activity unique to humans, and some scientists believe the advent of cooking played an important role in human evolution. Most anthropologists believe that cooking fires first developed around 250,000 years ago. The development of agriculture, commerce and transportation between civilizations in different regions offered cooks many new ingredients. New inventions and technologies, such as pottery for holding and boiling water, expanded cooking techniques. Some modern cooks apply advanced scientific techniques to food preparation.

There are very many methods of cooking, most of which have been known since antiquity. These include baking, roasting, frying, grilling, barbecuing, smoking, boiling, steaming and braising. A more recent innovation is microwaving. Various methods use differing levels of heat and moisture and vary in cooking time. The method chosen greatly affects the end result.

Although cooking has traditionally been a process carried out informally in a home or around a communal fire, cooking is also often carried out outside of personal quarters, for example at restaurants, or schools. Bakeries were one of the earliest forms of cooking outside the home, and bakeries in the past often offered the cooking of pots of food provided by their customers as an additional service. In the present day, factory food preparation has become common, with many "ready-to-eat" foods being prepared and cooked in factories and home cooks using a mixture of scratch made, and factory made foods together to make a meal.

Food Safety

When heat is used in the preparation of food, it can kill or inactivate potentially harmful organisms, such as bacteria and viruses, as well as various parasites such as tapeworms and Toxoplasma gondii. Food poisoning and other illness from uncooked or poorly-prepared food may be caused by bacteria such as pathogenic strains of Escherichia coli, Salmonella typhimurium and Campylobacter, viruses such as noroviruses, and protozoa such as Entamoeba histolytica. Parasites may be introduced through salad, meat that is uncooked or done rare, and unboiled water.

The sterilizing effect of cooking will depend on temperature, cooking time, and technique used. However, some bacteria such as Clostridium botulinum or Bacillus cereus, can form spores that survive cooking, which then germinate and regrow after the food has cooled. It is therefore recommended that cooked food should not be reheated more than once to avoid repeated growths that allow the bacteria to proliferate to dangerous level.

Cooking prevents many foodborne illnesses that would otherwise occur if the food was eaten raw. Cooking also increases the digestibility of some foods such as grains. Many foods, when raw, are inedible, and some are poisonous. For example kidney beans are toxic when raw or improperly cooked, due to the presence of phytohaemagglutinin which can be inactivated after cooking for at least ten minutes at 100 °C. Slow cooker however may not reach the desired temperature and cases of poisoning from red beans cooked in slow cooker have been reported.

Preparation, handling, and storage of food are other considerations in food safety. The temperature range from 41°F to 135 °F (5 °C to 57 °C) is the "Danger zone" where bacteria is likely to proliferate, food therefore should not be stored in this temperature range. Washing of hands and surfaces, and avoidance of cross-contamination are good practices in food safety. Food prepared on plastic cutting boards may be less likely to harbor bacteria than wooden ones, other research however suggested otherwise. Washing and sanitizing cutting boards is highly recommended, especially after use with raw meat, poultry, or seafood. Hot water and soap followed by a rinse with an diluted antibacterial cleaner, or a trip through a dishwasher with a "sanitize" cycle, are effective methods for reducing the risk of illness due to contaminated cooking implements.

Effects on Nutritional Content of Food

Proponents of Raw foodism argue that cooking food increases the risk of some of the detrimental effects on food or health. They point out that the cooking of vegetables and fruit containing vitamin C both elutes the vitamin into the cooking water and degrades the vitamin through oxidation. Peeling vegetables can also substantially reduce the vitamin C content, especially in the case of potatoes where most vitamin C is in the skin. However, research using an artificial gut has shown that in the specific case of carotenoids a greater proportion is absorbed from cooked vegetables than from raw vegetables.

German research in 2003 showed significant benefits in reducing breast cancer risk when large amounts of raw vegetable matter are included in the diet. The authors attribute some of this effect to heat-labile phytonutrients. Sulforaphane, which may be found in vegetables such as broccoli, has been shown to be protective against prostate cancer, however, much of it is destroyed when the vegetable is boiled.

Cooking and Carcinogens

In a human epidemiological analysis by Richard Doll and Richard Peto in 1981, diet was estimated to cause perhaps around 35% of cancers. Some of these cancers may be caused by carcinogens in food generated during cooking process, although it is often difficult to identify the specific components in diet that serve to increase cancer risk. Many food, such as beef steak and broccoli, contain low concentrations of both carcinogens and anticarcinogens.

Several studies published since 1990 indicate that cooking meat at high temperature creates heterocyclic amines (HCAs), which are thought to increase cancer risk in humans. Researchers at the National Cancer Institute found that human subjects who ate beef rare or medium-rare had less than one third the risk of stomach cancer than those who ate beef medium-well or well-done. While eating meat raw may be the only way to avoid HCAs fully, the National Cancer Institute states that cooking meat below 212 °F (100°C) creates "negligible amounts" of HCAs. Also, microwaving meat before cooking may reduce HCAs by 90%. Nitrosamines, present in processed and cooked foods, have also been noted as being carcinogenic, being linked to colon cancer.

Research has shown that grilling, barbecuing and smoking meat and fish increases levels of carcinogenic Polycyclic aromatic hydrocarbons (PAH). In Europe, grilled meat and smoked fish generally only contribute a small proportion of dietary PAH intake since they are a minor component of diet – most intake comes from cereals, oils and fats. However, in the US, grilled/barbecued meat is the second highest contributor of the mean daily intake of benzo[a]pyrene at 21% after 'bread, cereal and grain' at 29%.

Baking, grilling or broiling food, especially starchy foods, until a toasted crust is formed generates significant concentrations of acrylamide, a possible carcinogen.

REFERENCES

Jacob B M, *The Chemistry and Technology of Food and Food Products*, Inter science Publishers,New York.

NIIR Board Of Food Technologists, *Modern Technology of Food Processing & Agro Based Industries*, National Institute Of industrial Research, New Delhi.

Potter N and Hotchikiss J, *Food Science*, CBS Publishers & Distributors, New Delhi.

8

Fermented Cereal Foods

The global importance of cereal crops to the human diet and moreover to the written history of man and agriculture cannot be over stated. Cereal grains are the fruit of plants belonging to the grass family (*Gramineae*). The sustenance provided by cereals is frequently mentioned in the Bible, and they are by many other criterea the most important group of food crops produced in the world. Cereal crops are energy dense, containing 10 000-15 000 kJ/Kg, about 10-20 times more energy than most succulent fruits and vegetables. Nutritionally, they are important sources of dietary protein, carbohydrates, the B complex of vitamins, vitamin E, iron, trace minerals, and fiber. It has been estimated that global cereal consumption directly provides about 50 percent of protein and energy necessary for the human diet, with cereals providing an additional 25 percent of protein and energy via livestock intermediaries. Some cereals, notably wheat, contain proteins that form gluten, which is essential for making leavened bread. Although dried cereal grains constitute living cells that respire, when kept in an appropriate environment, whole grains can be stored for many years. In 1996, world cereal production amounted to more than two billion metric tons. Major cereal crops produced worldwide include wheat, rice, maize and barley. Other major cereal crops produced include sorghum, oats, millet and rye. Asia, America, and Europe produce more than 80 percent of the world's cereal grains. Wheat, rice, sorghum, and millet are produced in large quantities in Asia; corn and sorghum are principal crops in America, and barley, oats and rye are major crops in the former USSR and Europe.

Cereals have a variety of uses as food. Only two cereals, wheat and rye, are suited to the preparation of leavened bread. The most general usage of cereals is in cooking, either directly in the form of grain, flour, starch, or as semolina, etc. Another common usage of cereals is in the preparation of alcoholic drinks such as whiskey and beer (barley; sorghum), vodka (wheat), American bourbon (rye), Japanese sake (rice), etc. A variety of unique, indigenous fermented foods, other than leavened breads and alcoholic beverages, are also produced in regions of the world that rely mainly on plant sources of protein and calories. In developed countries, that obtain most of their protein from animal products, cereals are increasingly used as animal feed. More than 70 percent of the cereal crop produced in developed countries is fed to livestock; whereas, in developing countries, 68-98 percent of the cereal crop is used for human consumption.

The origin of farming practice appears to be located in the "Fertile Crescent", a wide belt of Southeast Asia, which includes Southern Turkey, Palestine, Lebanon and North Iraq. Abundant rainfall occured in the highlands of this area where there still exists a wide variety of wild cereals. *Triticum dicoccoides* (wheat) and *Hordeum spontaneum* (barley) were collected by local dwellers. There is evidence that the people of Uadi el-Natuf Tell of Southeast Asia were the first grain cultivators at about 7 800 B.C. By 5 000 BC, wild animals became rare forming only 5 percent of the diet, while cereals and farmed animals provided a sizeable part of human food. Early wild wheat (*Triticum*) and barley (*Hordeum*) species were diploid, carried few seeds, and abscissed from the plant at maturation, thus making harvest difficult. Polyploid plants can originate in nature but have little chance for self-propagation without cultivation. The cultivation and irrigation of cereals allowed the expansion of polyploid grains. The polyploid grains exhibit less genetic variation, since each gene is represented in several copies, causing more genetic uniformity and considerable increase in stability and yield. The first stable lines of polyploid cereals were identified as early as 6 000 BC. On the other hand, genetic variability in diploid wild types was essential in order to develop plants adapted to different environmental conditions and geographic areas. *Triticum turgide dicoccoides* was crossed with *Triticum fanschii* to give *Triticum aestvum*, the progenitor of actual wheat. *T. aestivum* has 42 chromosomes compared to the 14 chromosomes *of T. monococcum*. Today there are more than 20 000 cultivars of *T. aestivum* over the world. In Roman times, *T. dicoccoides* (spelt) was used for breadmaking and *T. vulgaris* (siligo) was used for soups because of it's low gluten content. Wheat and barley were initially cultivated in Southeast Asia, whereas rice was cultivated in Asia, maize in America, and sorghum and millet in Africa. Over the last 200 years active programmes in genetic selection and manipulation have changed the character of the original *Triticacee* from few grains and low gluten to abundant grains rich in gluten forming proteins. Triticale (genus *Trticosecale*) is a relatively new cereal that was developed by crossing wheat and rye in order to combine the tolerance of rye for poor soil and climatic conditions with the superior technological characteristics of wheat.

Cereals

Wheat

One of the oldest of all cutivated plants. Today, there are more than 50000 cultivars of wheat in existence and as a result wheat can be grown in a relatively wide range of climatic conditions. Growing best in temperate climates, it is susceptible to disease in warm, humid regions and cannot be grown as far from the equator as can rye and oats. Wheat was brought to America early in the seventeenth century where it came to prominence in the Great Plains by 1855. Different types of wheat are classified based on planting season and endosperm composition. Wheat holds a special place amongst the cereals because upon mixing wheat flour with water, an elastic matrix called "gluten" required for the production of leavened breads is formed. "Hard wheats" tend to contain relatively high levels of starch and relatively low levels of protein, while the reverse is true for "soft wheats". High protein flours are best suited for pastas and breads, while flour from soft wheats is excellent for cakes and pastries, etc.

Rice

The second most abundant cereal crop originated in the Indian subcontinent and Africa. Today, 90 percent of the world rice crop is grown in Asia. Alexander the Great is credited with introducing rice to Europe around 300 BC. Growing rice requires more water than other cereal crops, although rice is a highly productive crop. There are several thousand rice cultivars which may differ in color, aroma and grain size. The main commercial distinction between rice types is the grain size, i.e. long, medium and short. Long grain rice, also called "Indian", tends to separate relatively easily on cooking and is dry and flakey. Short grain rice, also called "Japanese" is sticky, moist and firm when cooked. Unlike wheat, rice is most often consumed as grain rather than as a flour. Different grades of milling include brown rice (hull removed), unpolished rice (hull, bran and most of germ removed), and polished rice (aleurone layer removed from unpolished rice). Since polishing removes most of the lipid, the latter product is relatively stable during storage. The discovery that rice bran can alleviate beriberi led to the discovery of the vitamin thiamine. The traditional technique of parboiling rice in India and Pakistan (also called "converted rice") prior to milling improves the nutritional quality of the grain by allowing the B vitamins in the bran and germ to diffuse into the endosperm.

Wild rice is native to the Great Lakes region of North America where it was originally harvested from the wild by native Indians. Although wild rice is now cultivated, it is expensive and accounts for less than 1 percent of the American rice market. The rice is first fermented to develop a nutty flavor and to ease hulling.

Maize

Corn was originally cultivated in Central America and became the staple of the Incas of Peru, the Mayas and Aztecs of Mexico and early cliff dwellers of the American Southwest. Columbus brought corn back to Europe where it became a popular crop in the south. Different types of maize are classified on the basis of their protein content and the hardness of the kernel. These include pop, flint, flour, Indian and sweet corns. Much of the niacin in corn is in a bound form and this led to pellagra in areas where corn became the food staple. It was not until the late 1920s that pellagra was identified as a vitamin deficiency. The traditional practice by early American natives, of boiling corn in 5 percent lime or ashes, releases bound niacin making it available as a nutrient.

Millet and Sorghum

Millet and sorghum are often grouped together because their growing conditions, processing and uses are similar. Millets are native to Africa or Asia and have been cultivated for more than 6 000 years. Millets grow well in arid regions with poor soils and are valued for their relatively high protein content among the cereals. Sorghum originated in East Africa and today is an important food crop in Africa, Asia, India and China where it is made into porridge, unleavened bread ("roti" in India) and beer. Whole grain sorghum flour has a relatively short shelf-life and production of low-fat sorghum flour requires removal of about 20 percent of the grain weight by abrasion. In North America, millet and sorghum are used primarily as livestock feed.

Barley

In the United States, barley is mostly used for feed, brewing and alcohol production with only about 2 percent used for human food. Barley flour is produced by abrasion dehulling, followed by milling of the "pearled" barley. Shellenberger, reviewed the uses of barley flour and grits in products such as soups, dressings, and baby foods.

Oats

About 95 percent of the world oat crop is used for livestock feed. However, oat consumption as human food has recently increased to 19 percent in the United States, perhaps due to the reported health benefits of the soluble fibre of oats. Oats thrive in a moist, cool climate and became an important crop in Northern Europe at the beginning of the seventeenth century. Oats have a relatively minor status among cereals because they are more difficult to process and are unstable due to their high lipid content and lipase activity. Reports of the possible blood cholesterol lowering effect of oat bran have increased the popularity of its use for human food in developed countries.

Rye

Rye appears to have originated in central Asia around 4 000 BC as a weed contaminating barley and wheat and became domesticated around the Baltic Sea in 400 BC where it grows well in a cool, moist climate with poor soil. It was traditionally a popular grain for bread making in Northern and Eastern Europe. Rye flour has a relatively low gluten content compared to wheat flour, but contains a unique class of carbohydrates (pentosans) that facilitate bread making. The milling of rye yields a flour that is classified based on color or ash content.

Botanical Structure of Cereals

All cereals belong to the taxonomic family known as the *Gramineae.* Other globally important crops in the *Gramineae* family include sugar cane and bamboo. Botanically speaking, cereal grains are a type of "dry" fruit called a caryopse. A botanical fruit is defined as the ripened ovary or the ovary and adjoining parts. The flesh of the fruit may originate from the floral receptacle, from carpillary tissue, or from extrafloral structures, such as bracts. The caryopse fruit structure differs from that of other fruits (e.g., fleshy fruits) in that a thin, dry fruit wall is fused together with the seed coat. Cereals are sometimes thought of as seeds by the layman since the bulk of their tissue is the true seed; i.e. the part resulting from sexual reproduction of the plant. Seeds are produced within the fruit which serves to protect and aid their dispersal for propagation in a variety of ways for different plants. The seed is the result of sexual reproduction of the plant when the flower is fertilized and is the organ of propagation. Kernel structure is important with respect to minimizing damage during grain harvest, drying, handling, storage, milling, germination and in enhancing nutritional value. Selection of cultivars with a large embryo or without hulls for example, will improve the protein content if the entire grain is consumed.

The seed portion of cereals consists of numerous components which basically include three parts: a seed coat or testa (bran), storage organ or nutritive reserve for the seed (endosperm), and a miniature plant or germ. The fruit tissue consists of a layer of epidermis and several thin inner layers a few cells thick. The aleurone layer which is just below the seed coat, is only a few cells thick, but is rich in oil, minerals, protein and vitamins. Starch and protein are located in the endosperm which represents the bulk of the grain and is sometimes the only part of the cereal consumed. Starch is housed in the form of subcellular structures called granules that are embedded in a matrix of protein. The developing endosperm contains protein bodies which become a continous phase as the grain matures. There is generally a gradient of more protein and less starch per cell from the outer to the inner region of the endosperm. The diameter, shape, size distribution and other characteristics of starch granules vary with different cereals. Starch granules range in size from 3-8 μm in rice; 2-30 μm in corn, and 2-55 μm in wheat. Reserve proteins in the endosperm are in the form of smaller "protein bodies" that range in size from 2-6 μm that become disordered and adhere to the starch granules in the mature grain of species like wheat.

Nutritional Quality of Cereals

Cereals, together with oil seeds and legumes, supply a majority of the dietary protein, calories, vitamins, and minerals to the bulk of populations in developing nations. Cereal grains are low in total protein compared to legumes and oilseeds. Lysine is the first limiting essential amino acid for man; although rice, oats and barley contain more lysine than other cereals. Corn protein is also limiting in the essential amino acid tryptophan, while other cereals are often limiting in threonine. The annual global yield of essential amino acids from major cereals has been compared to a hypothetical population of 3 billion adults and 2 billion children. Accordingly, if all cereals were effectively and fully utilized for human consumption they would more than meet man's needs for essential amino acids.

Barley, sorghum, rye and oat proteins have lower digestibilities (77-88%) than those of rice, maize and wheat (95-100%). The biological value and net protein utilization of cereal proteins is relatively low due to deficiencies in essential amino acids and low protein availability. The digestible energy of rice is significantly better than that of other cereals.

Cereals also provide B-group vitamins and minerals, although refining results in losses of these nutrients. The endosperm of wheat contains only about 0.3% ash. Phosphorous, potassium, magnesium, calcium and traces of iron and other minerals are found in cereals. Barley and wheat provide 50 and 36 mg Ca/100 g respectively. Barley provides 6 mg of iron per 100 g; millet provides 6.8; oats, 4.6 and wheat, 3.1. In contrast, soybeans provide more of these nutrients, i.e., Ca (210 mg/100 g) and Fe (7 mg/100 g). Some grains, notably barley, sorghum, and oats, contain appreciable amounts of crude fiber and are referred to as coarse grains. The nutritive and sensory value of cereal grains and their products are, for the most part, inferior to animal food products. Methods that can be employed to improve the nutritive value of cereals include traditional genetic selection, genetic engineering, amino acid and other nutrient fortification, complementaion with other proteins (notably legumes), milling, heating, germination and fermentation.

Antinutrients and Toxic Components in Cereals

Cereals and other plant foods may contain significant amounts of toxic or antinutritional substances. In this regard, legumes are a particularly rich source of natural toxicants including protease inhibitors, amylase inhibitors, metal chelates, flatus factors, hemagglutinins, saponins, cyanogens, lathyrogens, tannins, allergens, acetylenic furan and isoflavonoid phytoalexins. Most cereals contain appreciable amounts of phytates, enzyme inhibitors, and some cereals like sorghum and millet contain large amounts of polyphenols and tannins. Some of these substances reduce the nutritional value of foods by interfering with mineral bioavailability, and digestibility of proteins and carbohydrates. Since legumes are often consumed together with cereals, proper processing of cereal-legume mixtures should eliminate these antinutrients before consumption. Relatively little is known about the fate of antinutrients and toxicants in traditional fermented foods.

Phytates

Phytic acid is the 1,2,3,4,5,6-hexaphosphate of myoinositol that occurs in discrete regions of cereal grains and accounts for as much as 85% of the total phosphorous content of these grains. Phytate reduces the bioavailability of minerals, and the solubility, functionality and digestibility of proteins and carbohydrates. Fermentation of cereals reduces phytate content via the action of phytases that catalyze conversion of phytate to inorganic orthophosphate and a series of myoinositols, lower phosphoric esters of phytate. A 3-phytase appears to be characteristic of microorganisms, while a 6-phytase is found in cereal grains and other plant seeds.

Tannins

Oligomers of flavan-3-ols and flavan-3,4-diols, called condensed tannins, occur widely in cereals and legumes. These compounds are concentrated in the bran fraction of cereals. Tannin-protein complexes can cause inactivation of digestive enzymes and reduce protein digestibility by interaction of protein substrate with ionizable iron. The presence of tannins in food can therefore lower feed efficiency, depress growth, decrease iron absorption, damage the mucosal lining of the gastrointestinal tract, alter excretion of cations, and increase excretion of proteins and essential amino acids. Dehulling, cooking and fermentation reduce the tannin content of cereals and other foods.

Saponins

These sterol or triterpene glycosides occur widely in cereals and legumes. Saponins are detected by their hemolytic activity and surface active properties. Although the notion that they are detrimental to human health has been questioned, they have been reported to cause growth inhibition.

Enzyme Inhibitors

Protease and amylase inhibitors are widely occurent in seed tissues including cereal grains. Trypsin-, chymotrypsin-, subtilisin-inhibitor, and cysteine-protease inhibitors are present in all major rice cultivars grown in California, although the individual inhibitor amounts are quite varaiable and are concentrated in the bran fraction. They are believed to cause growth inhibition by interfering with digestion, causing pancreatic hypertrophy and metabolic disturbance of sulfur amino acid utilization. Although these inhibitors tend to be heat stable, there are numerous reports that trypsin inhibitor, chymotrypsin inhibitor, and amylase inhibitor levels are reduced during fermentation.

Fermented Cereals

Animal or plant tissues subjected to the action of microorganisms and/or enzymes to give desirable biochemical changes and significant modification of food quality are referred to as fermented foods. Fermentation is the oldest known form of food biotechnology; records of barley conversion to beer date back more than 5000 years. According to Steinkraus, the traditional fermentation of foods serves several functions:

1. Enrichment of the diet through development of a diversity of flavors, aromas, and textures in food substrates
2. Preservation of substantial amounts of food through lactic acid, alcoholic, acetic acid, and alkaline fermentations
3. Enrichment of food substrates biologically with protein, essential amino acids, essential fatty acids, and vitamins
4. Detoxification during food fermentation processing
5. A decrease in cooking times and fuel requirements.

Aside from alcoholic fermentations and the production of yogurt and leavened bread, food fermentations continue to be important primarily in developing countries where the lack of resources limits the use of techniques such as vitamin enrichment of foods, and the use of energy and capital intensive processes for food preservation. The technology of producing many indigenous fermented foods from cereals remains a household art in these countries. Prospects for applying advanced technologies to indigenous fermented foods and for the production of value-added additive products, such as colors, flavors, enzymes, antimicrobials, and health products during food fermentations have been reviewed.

Special mention should be made of the microbiological risk factors associated with fermented foods. The safety of fermented foods has been recently reviewed. Cases of food-born infection, and intoxications due to microbial metabolites such as mycotoxins, ethyl carbamate, and biogenic amines have been reported in fermented foods. Major risk factors include the use of contaminated raw materials, lack of pasteurization, and use of poorly controlled fermentation conditions. On the other hand, non-toxigenic microorganisms can serve to antagonize pathogenic microorganisms and even degrade toxic substances such as mycotoxins in fermented foods.

Indigenous Fermented Cereal Foods

Most bacterial fermentations produce lactic acids; while yeast fermentation results in alcohol production. Many of the indigenous fermentation products of cereals are valued for the taste and aroma active components produced and are used as seasonings and condiments. A summary of flavor compounds formed in such products was compiled by Chaven and Kadam. A number of fermented products utilize cereals in combination with legumes, thus improving the overall protein quality of the fermented product. Cereals are deficient in lysine, but are rich in cystine and methionine. Legumes on the other hand are rich in lysine but deficient in sulfur containing amino acids. Thus, by combining cereals with legumes, the overall protein quality is improved. The Chinese concept of "fan" (rice) and "tsai"(other vegetables) for a balanced and interesting diet is seen throughout the world.

Importance and Benefits of Fermented Cereals

Fermented foods contribute to about one-third of the diet worldwide. Cereals are particularly important substrates for fermented foods in all parts of the world and are staples in the Indian subcontinent, in Asia, and in Africa. Fermentation causes changes in food quality indices including texture, flavor, appearance, nutrition and safety. The benefits of fermentation may include improvement in palatability and acceptability by developing improved flavours and textures; preservation through formation of acidulants, alcohol, and antibacterial compounds; enrichment of nutritive content by microbial synthesis of essential nutrients and improving digestibility of protein and carbohydrates; removal of antinutrients, natural toxicants and mycotoxins; and decreased cooking times.

The content and quality of cereal proteins may be improved by fermentation. Natural fermentation of cereals increases their relative nutritive value and available lysine. Bacterial fermentations involving proteolytic activity are expected to increase the biological availability of essential amino acids more so than yeast fermentations which mainly degrade carbohydrates. Starch and fiber tend to decrease during fermentation of cereals. Although it would not be expected that fermentation would alter the mineral content of the product, the hydrolysis of chelating agents such as phytic acid during fermentation, improves the bioavailability of minerals. Changes in the vitamin content of cereals with fermentation vary according to the fermentation process, and the raw material used in the fermentation. B group vitamins generally show an increase on fermentation. During the fermentation of maize or kaffircorn in the preparation of kaffir beer, thiamine levels are virtually unchanged, but riboflavin and niacin contents almost double.

Reddy and Pierson, reviewed the effect of fermentation on antinutritional and toxic components in plant foods. Fermentation of corn meal and soybean-corn meal blends lowers flatus producing carbohydrates, trypsin inhibitor and phytates. However, fermentation of cereals with fungi, such as *Rhizopus oligosporus*, has been reported to release bound trypsin inhibitor, thus increasing it's activity. Fungal and lactic acid fermentations have also been reported to reduce aflatoxin B1, sometimes by opening of the lactone ring which results in complete detoxification.

Another benefit of fermentation is that frequently the product does not require cooking or the heating time required for preparation is greatly reduced.

Need for Additional Research

Some advantages of traditional fermentations are that they are labor-intensive, integrated into village life, familiar, utilize locally produced raw materials, inexpensive, have barter potential and the subtle variations resulting, add interest and tradition to local consumers. From this perspective, research leading to new fermentation technologies should be sensitive to social and economic factors in developing countries. Rapid displacement of traditional foodstuffs in developing countries with technologies developed in more affluent countries may result in centralised production, distribution problems, less local involvement in food processing, less employment in some areas, less nutritionally adequate substitutions in raw materials, displacement of traditional arts, loss of unique local know-how, dependence on importation of equipment and materials, initially require the use of outside consultants, and may otherwise not meet local needs as fully as traditional fermented products. On the other hand, indigenous fermentations may have a number of problems, i.e., they are uncontrolled and often unhygenic, labor intensive, seen as primitive by some people, are normally not integrated into the economic mainstream, difficult to tax, have limited export potential and in some cases, the impact on nutritive value and safety is questionable.

Specific microflora involved with indigenous fermentations is, in many cases, not known at this time. Specific information on microflora appears to be lacking for several indigenous fermented cereal products. The microbiology of many of these fermentations is undoubtedly quite complex. Many indigenous cereal fermentations involve the combined action of bacteria, yeast and fungi. Some microflora may participate in parallel while others may participate in a sequential manner with a changing dominant flora during the course of the fermentation. The specific microflora involved may vary somewhat from village to village and from family to family within the same village. The identification of specific microflora involved is needed to amplify and control such positive factors as the excretion of lysine by strains of *Lactobacillus plantarum* and the metabolic detoxification of mycotoxins by *Rhizopus oryzae*; as well as to minimize or prevent negative factors such as growth and metabolism of pathogenic and toxinogenic bacteria, e.g., bongrek acid and toxoflavin formation by *Pseudomonas cocovenenans*. Identifying and providing a practical means of using appropriate starter cultures is advantageous due to the competitive role of microorganisms and their metabolites in preventing growth and metabolism of unwanted microorganisms. A strong starter may reduce fermentation times, minimise dry matter losses, avoid contamination with pathogenic and toxigenic bacteria and molds, and minimize the risk of incidental microflora causing off-flavor, etc. According to Nout optimization of starter cultures may be achieved by either conventional selection and mutation, or by recombinant-DNA techniques to result in increased levels of safety. Relatively litle is known of the contribution of microflora to the formation of desired flavor notes during such fermentations. Genes for flavor and other beneficial enzymes that come from incidental microflora may be incorporated into starter bacteria to facilitate more subtle and ancillary aspects of the fermentation along with primary events such as lactic acid production, thus preserving the distinctive nature of products made in different regions.

The contribution of specific enzymes to indigenous cereal fermentations is perhaps even less understood than that of microorganisms. It is likely that there is considerable synergy between complimentary enzymes from the cereal itself and from the microorganisms. One known example of this is the reduction of phytates resulting from 6-phytases of cereal origin and 3-phytases of microbial origin. Another is the synergy of cereal enzymes and yeast in bread making. It is interesting that fermentation of cereals (e.g. breadmaking and brewing) in the Western world was adversely affected in some ways by the introduction of modern dehydration and storage techniques that minimized fungal contamination and incipient germination. Partial germination of cereal in the field and contamination with otherwise innocuous fungal contaminants contribute enzymes, notably a-amylase and proteases, that aid these fermentations. Today, essentially all beer production and continuous breadmaking in the West is achieved with the aid of added enzymes. A similar situation may occur in developing countries, i.e. as improvements in cereal handling are introduced to minimise postharvest losses and mycotoxin formation, the otherwise improved crop may be less suitable in some ways for traditional fermentations. Hence, basic information is need on the contribution of cereal enzymes and other constituents to indigenous fermentations. With this information in hand, consideration can then be given to use of enzyme supplements and other additives to improve the rate and quality of fermentations.

Another consideration for future research is the contribution of the aforementioned enzyme inhibitors in cereal fermentations. In addition to their already discussed significance as antinutrients in the finished product; protease, amylase and other enzyme inhibitors are expected to influence the rate and extent of important bioconversions that occur during indigenous fermentations. The concentration and spectrum of enzyme inhibitors varies considerably between cereal cultivars. Not withstanding the benefits of the continual introduction of new cereal varieties, given the genes for enzyme inhibitors are part of the defensive system of plants against insects and other pests, it is possible that introduction of "improved" cereal cultivars in developing countries may adversely affect the utility of cereals for indigenous fermentation. For this reason, basic research on the participation of cereal enzyme inhibitors in the process may provide useful insights on the need for including tests for inhibitors prior to introducing new varieties in areas that extensively utilise cereal fermentations to produce staple foodstuffs.

As pointed out by Wood, there is a possible backlash if consumers in developing countries abandon traditional fermented foods for "smart," sophisticated products popularized in Europe and America. For example, the replacement of indigenous fermented cereal drinks with cola beverages could have a significant negative impact on daily nutrition of many consumers in developing countries. Study of traditional fermentations will undoubtedly yield new information that will expand our global knowledge of science and impact technology throughout the world. Thus, basic research of indigenous cereal fermentations will lead to "inward" as well as "outward" technology transfer.

References

Bowers, J. 1992. *Food Theory and Applications*. New York, Macmillan Publishing Co., 777 pp.

Campbell-Platt, G. 1994. "Fermented foods- a world perspective." *Food Research International* 27: 253.

Cook, P. E. 1994. "Fermented foods as biotechnological resources." *Food Research International* 27: 309.

Nout, M. J. R. 1994. "Fermented foods and food safety." *Food Research International* 27: 291.

Steinkraus, K. H., Ed. 1995. *Handbook of Indigenous Fermented Foods*. New York, Marcel Dekker, Inc., 776 pp.

Bibliography

Alais, C. and G. Linden. 1991. *Food Biochemistry*. New York, Ellis Horwood Ltd., 222 pp.

Atkins, P.J., 2000. "The pasteurization of England: the science, culture and health implications of milk processing, 1900-1950", In: Smith, D. & Phillips, *Journal of Food, science and regulation in the 20th century*, Routledge,

Betschart, A. A. 1982. "World food and nutrition problems." *Cereal Food World* 27: 562.

Biaugeaud, H. 1994. Food processing equipment. *Technical Bulletin*. Henri Biaugeaud, S.A.: Arcueil.

Borgstrom, G. 1968. *Principals of Food Science*, Vol. 2. Food Microbiology and Biochemistry. New York, Macmillan.

Bowers, J. 1992. *Food Theory and Applications*. New York, Macmillan Publishing Co., 777 pp.

Campbell-Platt, G. 1994. "Fermented foods- a world perspective." *Food Research International* 27: 253.

Colin, D. 1992. Recent trends in fruit and vegetable processing. In *Food Science and Technology Today*, 7, (2)), pp. 111 - 116

Cook, P. E. 1994. "Fermented foods as biotechnological resources." *Food Research International* 27: 309.

FAO. 1990. *Rural processing and preserving techniques for fruits and vegetables*. Rome: FAO.

Fellows, P., 2000. *Food Processing Technology, Principles and Practice*, 2nd Edition, Woodhead, Cambridge.

Gegner, L., 2001."Value-Added Dairy Options", *ATTRA* (Appropriate Technology Transfer for Rural Areas), Fayetteville, Ark., Aug.

Good, A. 1997. Fruit factory in the forest, *Food Chain* 21, 4-5.

Hanke, H. 1992. The use of fruit juices in confectionery products.In 46th P.M.C.A. Production Conference, 1992.

Herklots, J. 1991. Making money from honey, *Food Chain,* 3, 3-5.

Hidellage, V. 1999. Empowering small-scale cashew processors in Sri Lanka, *Food Chain,* 24, 11-15.

Hinman, Robert B., Harris, Robert B. *The Story of Meat.* Swift & Company, 1939. Katherine.

Jacob B M, *The Chemistry and Technology of Food and Food Products*, Inter science Publishers,New York.

Jayaraj, J. 1999. Training in food processing - a sustainable approach in India, *Food Chain,* 24, 19-21.

Karmas, E. 1976. *Processed meat technology*. New Jersey, USA, Noyes Data Corporation.

NIIR Board Of Food Technologists, *Modern Technology of Food Processing & Agro Based Industries*, National Institute Of industrial Research, New Delhi.

Nout, M. J. R. 1994. "Fermented foods and food safety." *Food Research International* 27: 291.

Potter N and Hotchikiss J, *Food Science*, CBS Publishers & Distributors, New Delhi.

Pottter, N.N. 1984. *Food Science.* 3rd Edition. AVI Publishing Co. Westport, Conn.

Robinson, R. K., 1986. *Modern Dairy Technology*, Volume 1. Elsevier AppliedScience Publishers, London.

Steinkraus, K. H., Ed. 1995. *Handbook of Indigenous Fermented Foods*. New York, Marcel Dekker, Inc., 776 pp.

Toledo, R.T., 1991. *Fundamentals of Food Process Engineering*, 2nd Edition, van Nostrand, Reinhold, New York.

Torregiani, D. 1993. Osmotic dehydration in fruit and vegetable processing. In *Food Research International.*

Trager, J. 1996. *The Food Chronology,* Aurum Press, London, UK

Tucker, G.A. and L.F.J. Woods. 1995. *Enzymes in Food Processing.*Chapman and Hall, New York, 319 pp.

Varnam, A. H. and Sutherland, J. P., 1997. *Milk and Milk Products:Technology, Chemistry and Microbiology*, Chapman and Hall, London, World Bank.

Winrock International. 1992. *Animal agriculture in developing countries: technology dimensions*. Morrilton, AR, USA, Winrock International.